DNA MODIFICATIONS IN THE BRAIN

DNA MODIFICATIONS IN THE BRAIN

Neuroepigenetic Regulation of Gene Expression

Edited by

TIMOTHY W. BREDY

Department of Neurobiology and Behavior
The Francisco J. Ayala School of Biological Sciences
Bonney Research Laboratory
University of California Irvine
Irvine, CA, United States

Amsterdam • Boston • Heidelberg • London
New York • Oxford • Paris • San Diego
San Francisco • Singapore • Sydney • Tokyo
Academic Press is an imprint of Elsevier

Academic Press is an imprint of Elsevier
125 London Wall, London EC2Y 5AS, United Kingdom
525 B Street, Suite 1800, San Diego, CA 92101-4495, United States
50 Hampshire Street, 5th Floor, Cambridge, MA 02139, United States
The Boulevard, Langford Lane, Kidlington, Oxford OX5 1GB, United Kingdom

Library of Congress Cataloging-in-Publication Data
A catalog record for this book is available from the Library of Congress

British Library Cataloguing-in-Publication Data
A catalogue record for this book is available from the British Library

ISBN: 978-0-12-801596-4

For information on all Academic Press publications
visit our website at https://www.elsevier.com/

Working together
to grow libraries in
developing countries

www.elsevier.com • www.bookaid.org

Publisher: Mara Conner
Acquisition Editor: Natalie Farra
Editorial Project Manager: Kathy Padilla
Production Project Manager: Karen East and Kirsty Halterman
Designer: Matthew Limbert

Typeset by TNQ Books and Journals

CONTENTS

LIST OF CONTRIBUTORS

V.V. Ashapkin
Lomonosov Moscow State University, Moscow, Russia

T.W. Bredy
The University of California Irvine, Irvine, CA, United States; The University of Queensland, Brisbane, QLD, Australia

J.J. Day
University of Alabama at Birmingham, Birmingham, AL, United States

M. Fasolino
University of Pennsylvania, Philadelphia, PA, United States

J. Feng
Icahn School of Medicine at Mount Sinai, New York, NY, United States

P. Jin
Emory University, Atlanta, GA, United States

Y. Kang
Emory University, Atlanta, GA, United States

J. Korlach
Pacific Biosciences, Menlo Park, CA, United States

R. Lister
The University of Western Australia, Perth, WA, Australia; The Harry Perkins Institute of Medical Research, Perth, WA, Australia

X. Li
University of California Irvine, Irvine, CA, United States

P.R. Marshall
The University of California Irvine, Irvine, CA, United States

S. Morishita
The University of Tokyo, Tokyo, Japan

E.A. Mukamel
University of California San Diego, La Jolla, CA, United States

E.J. Nestler
Icahn School of Medicine at Mount Sinai, New York, NY, United States

Y. Suzuki
The University of Tokyo, Tokyo, Japan

B.F. Vanyushin
Lomonosov Moscow State University, Moscow, Russia

Z. Wang
Emory University, Atlanta, GA, United States

W. Wei
The University of Queensland, St Lucia, QLD, Australia

S.A. Welsh
University of Pennsylvania, Philadelphia, PA, United States

Z. Zhou
University of Pennsylvania, Philadelphia, PA, United States

PREFACE

The field of neuroepigenetics has a long and rich history, beginning with the discovery of experience-induced DNA modifications in the brain and other landmark observations over the past 40 years by Vanyushin and Ashapkin (Chapter 1), and extended by the recent discovery of downstream oxidative derivatives of 5-methylcytosine and the elucidation of their functional roles in brain development as investigated and discussed by Li and Wei (Chapter 2), Kang et al. (Chapter 3), and Fasolino et al. (Chapter 4). This information has led to the establishment of links between DNA modification and cognition and behavior related to neuropsychiatric disease described by Mukamel and Lister (Chapter 5) and Day (Chapter 6). Therefore, it has been unequivocally demonstrated that DNA modifications are dynamic and reversible across the life span and that they play an important role in the regulation of gene expression in both the normal and the diseased brain. Together with new insights regarding the mitochondrial neuroepigenome, as introduced by Suzuki et al. (Chapter 7), and the application of recent technical advances in DNA sequencing, as discussed by Feng and Nestler (Chapter 8), these new lines of research represent the leading edge in the quest to understand gene–environment interactions and how they influence the neuroepigenetic regulation of gene expression and its impact on subsequent behavioral adaptation. It is a remarkable time for neuroscience. As discussed by Marshall and Bredy (Chapter 9), armed with new technology and freedom from the constraints of dogma, we embark on entirely new directions in the study of DNA modifications in the brain. The work described herein serves to usher in this exciting new era.

Timothy W. Bredy

CHAPTER 1

History and Modern View on DNA Modifications in the Brain

B.F. Vanyushin, V.V. Ashapkin
Lomonosov Moscow State University, Moscow, Russia

INTRODUCTION

Almost 70 years ago it was discovered that, along with four classical bases, so-called "minor" bases are present in DNA. 5-Methylcytosine (5mC) was found first as a minor base in various DNAs (Hotchkiss, 1948; Wyatt, 1950), and N^6-methyladenine (m6A) was soon identified in bacterial DNA (Dunn & Smith, 1955). It was later found that mammalian DNA may also contain N^2-methylguanine and 3-methylcytosine (Culp, Dore, & Brown, 1970). The mechanism underlying the accumulation of these bases in DNA was unknown for a long period. Only in 1963 were the specific DNA methyltransferases first observed in bacteria (Gold & Hurwitz, 1963) and then in eukaryotes; these enzymes transferred methyl groups from S-adenosyl-L-methionine selectively onto definite cytosine or adenine residues in DNA chains. It became clear that minor bases (5mC and m6A) do not incorporate into DNA during synthesis, but they accumulate as a result of enzymatic modification (methylation) of common bases (C or A, respectively) in DNA chains that are either forming or already formed. Nevertheless, the specificity and functional role of DNA methylation remained unknown for a long time. Moreover, the concept that these minor bases do not have any essential significance both in the structure of DNA itself and its functioning was quite widely disseminated. The classic model system in traditional genetics, *Drosophila*, served mistakenly very often as "irrefutable" evidence for this postulate. In fact, 5mC in the fly genome escaped detection for a very long time, leading to the conclusion that this DNA modification does not play a significant role in eukaryotic organisms. This situation did not bring very much enthusiasm to DNA methylation research in many world-renowned molecular biology laboratories, which allowed us to study this particular epigenetic mechanism without competition for many years (Table 1.1).

Actually, we have been involved in this research for more than 50 years. Similar to the great Russian physiologist (Nobel Prize Laureate) Ivan Pavlov, who erected a memorial to the dog (his beloved experimental animal), we have to erect a memorial to *Drosophila* because the preceding situation with it allowed us to peacefully

DNA Modifications in the Brain
ISBN 978-0-12-801596-4
http://dx.doi.org/10.1016/B978-0-12-801596-4.00001-0

Table 1.1 Time line of the landmark discoveries

Years	Discovery	References
1948–50	5-Methylcytosine (5mC) found as a minor base in various DNAs	Hotchkiss (1948) and Wyatt (1950)
1955	N^6-Methyladenine (m6A) identified in bacterial DNA	Dunn and Smith (1955)
1959	High content of 5mC in plant DNA found	Vanyushin and Belozersky (1959)
1962–68	CpG dinucleotides shown to be the main target of DNA methylation in eukaryotes	Doskočil and Šorm (1962) and Grippo, Iaccarino, Parisi, and Scarano (1968)
1963	DNA methyltransferases observed in bacteria	Gold and Hurwitz, (1963)
1967–73	Species, tissue, and age specificities of DNA methylation in animals discovered; role of DNA methylation in gene expression, cell differentiation, and aging proposed	Berdyshev, Korotaev, Boyarskikh, and Vanyushin (1967), Vanyushin, Tkacheva, and Belozersky (1970), Vanyushin, Mazin, Vasilyev, and Belozersky (1973), and Vanyushin, Nemirovsky, Klimenko, Vasiliev, and Belozersky (1973)
1970	First experimental evidence that methylation affects double helical structure of DNA	Vanyushin, Belyaeva, Kokurina, Stelmashchyuk, and Tikhonenko (1970)
1974–77	First evidence that neuronal DNA methylation plays a role in learning	Vanyushin, Tushmalova, and Guskova (1974), Vanyushin, Tushmalova, Guskova, Demidkina, and Nikandrova (1977), and Guskova, Burtseva, Tushmalova, and Vaniushin (1977)
1975	Conception of the maintenance DNA methylation with methyltransferases acting on hemimethylated sites proposed	Riggs (1975) and Holliday and Pugh (1975)
1978	First use of the isoschizomeric restriction endonuclease pair HpaII and MspI to test for the presence of 5mC in individual gene sequence	Waalwijk and Flavell (1978)
1978–80	First direct indications on reverse correlation between DNA methylation and gene activity	Waalwijk and Flavell (1978) and Sutter and Doerfler (1980)
1980	Methylation of Okazaki fragments and longer replicative intermediates found to occur immediately after replication in plants and animals	Bashkite, Kirnos, Kiryanov, Aleksandrushkina, and Vanyushin (1980) and Kiryanov, Kirnos, Demidkina, Alexandrushkina, and Vanyushin (1980)
1981	Non-CpG methylation of DNA in plants discovered	Kirnos, Aleksandrushkina, and Vanyushin (1981) and Gruenbaum, Naveh-Many, Cedar, and Razin (1981)
1981–82	Clonal inheritance of DNA methylation patterns in dividing cells experimentally demonstrated	Wigler, Levy, and Perucho (1981) and Stein, Gruenbaum, Pollack, Razin, and Cedar (1982)
1983	Selective synthesis of DNA in rat brain induced by learning discovered; mechanism of brain DNA demethylation after learning by an excision repair mechanism proposed	Ashapkin, Romanov, Tushmalova, and Vanyushin (1983)

Year	Event	Reference
1986–90	Active demethylation of DNA discovered	Razin et al. (1986) and Paroush, Keshet, Yisraeli, and Cedar (1990)
1988–92	First mammalian DNA methyltransferases cloned, murine Dnmt1 and human DNMT1	Bestor, Laudano, Mattaliano, and Ingram (1988) and Yen et al. (1992)
1990–95	Existence of non-CpG methylation of DNA in animals firmly established	Toth, Mueller, and Doerfler (1990) and Clark, Harrison, and Frommer (1995)
1992	Targeted mutation of *Dnmt1* shown to cause severe developmental abnormalities and embryo lethality in mice	Li, Bestor, and Jaenisch (1992)
1998	First mammalian de novo DNA methyltransferases Dnmt3a and Dnmt3b cloned	Okano, Xie, and Li (1998b)
2000	5mC found in *Drosophila* DNA	Gowher, Leismann, and Jeltsch (2000) and Lyko, Ramsahoye, and Jaenisch (2000)
2000	High non-CpG methylation in embryonic stem cells discovered	Ramsahoye et al. (2000)
2001	Dnmt1 requirement for survival of mitotic neuronal precursors but not postmitotic neurons demonstrated by neuron-specific knockout *Dnmt1* mutation in mice	Fan et al. (2001)
2003	Activity-dependent demethylation of plasticity-related gene *Bdnf* in postmitotic neurons demonstrated	Martinowich et al. (2003)
2004–08	Early life experience shown to alter methylation status of genes in brain DNA	Weaver et al. (2004), Champagne et al. (2006), and Mueller and Bale (2008)
2007–08	Involvement of DNA methylation in learning rediscovered	Miller and Sweatt (2007) and Lubin, Roth, and Sweatt (2008)
2007–08	Gadd45a-assisted DNA demethylation by an nucleotide excision repair mechanism discovered	Barreto et al. (2007) and Rai et al. (2008)
2009	5-Hydroxymethylcytosine (5hmC) as a product of 5mC hydroxylation by the TET family oxygenases in animal DNA discovered	Kriaucionis and Heintz (2009) and Tahiliani et al. (2009)
2010	Dnmt1 and Dnmt3a requirement for neurogenesis and for learning and memory in mice demonstrated by *Dnmt1* and *Dnmt3a* knockouts	Feng et al. (2010) and Wu et al. (2010)
2012–13	Considerable levels of non–CpG methylation found in brain of mice and humans	Xie et al. (2012) and Varley et al. (2013)
2015	m6A found in *Drosophila melanogaster* DNA	Zhang et al. (2015)

investigate DNA methylation starting from the early beginnings without being tired out by enormous competition and agiotage. Besides, a long time ago we noted that the *Drosophila* genome is very much deficient in CpG sequences that usually serve as the main substrates for in vivo DNA methylation in eukaryotes; according to our opinion this strong CpG suppression in *Drosophila* genome could be due only to methylation of cytosine residues associated with deamination of 5mC (Mazin & Vanyushin, 1988). As we could not detect the proper DNA methyltransferase activity in *Drosophila* at that time, we designated this putative DNA modification as a "fossil" DNA methylation (Mazin & Vanyushin, 1988). Later, it was shown that DNA in *Drosophila* contains 5mC, with this DNA modification being important for normal insect development, and specific cytosine DNA methyltransferases have been detected at the early insect developmental stages (Gowher et al., 2000; Lyko et al., 2000). Furthermore, m6A has been found in *Drosophila* DNA (Zhang et al., 2015). Based on our findings in plants, we were always sure that these, and other, enzymatic genome modifications should not be superfluous in the genome organization and must have some function in the cell.

DNA METHYLATION AND ITS INFLUENCE ON DNA STRUCTURE AND INTERACTION WITH PROTEINS

We have been lucky to find unusual natural double-stranded DNA in AR9 bacteriophage of *Bacillus brevis* in which thymine is completely substituted by a typical RNA base, uracil. Basically, uracil is thymine lacking a methyl group. This bacteriophage DNA melted at significantly lower temperature compared with normal thymine-containing DNA of the equivalent base composition (Vanyushin, Belyaeva, et al., 1970). It became clear that methylation of cytosine residues is not indifferent to DNA structure: it stabilizes the double helix. Methylation of cytosine introduces a methyl group into an exposed position in the major groove of the DNA helix, and the binding of various proteins could be affected by such change (Razin & Riggs, 1980). It was well known that 5mC profoundly affects the binding of lac repressor to *lac* operator sequences, as well as the binding of bacterial restriction endonucleases to their recognition sites. The only question was whether eukaryotic cells use this mechanism to control regulatory protein binding to DNA. We have found a plant protein that binds specifically to regulatory elements of ribosomal RNA genes and showed that its binding is inhibited by in vitro methylation of cytosine residues in CCGG sites (Ashapkin, Antoniv, & Vanyushin, 1995). In many cases cytosine DNA methylation prohibits binding of specific nuclear proteins involved in transcription and other genetic processes. Conversely, there are proteins that bind specifically to methylated DNA sequences and arrange on DNA an entire ensemble of proteins controlling gene expression.

REPLICATIVE DNA METHYLATION AND THE INHERITANCE OF THE DNA METHYLATION PATTERN

Riggs (1975) and Holliday and Pugh (1975) proposed models in which symmetrical methylation of both DNA strands, coupled with a methyltransferase acting only on hemimethylated sites (now widely referred to as maintenance methyltransferase), would lead to stable maintenance of DNA methylation patterns through DNA replication. The methylated patterns of CpG-containing sites were indeed clonally inherited in dividing mouse cells, with a fidelity ~95% per cell generation (Stein et al., 1982; Wigler et al., 1981). We found that DNA synthesis in cells grown in a culture at high cell density pauses at the stage when most short DNA fragments in the lagging strand (Okazaki fragments) are still not ligated (Bashkite et al., 1980; Kiryanov et al., 1980). It turned out that Okazaki fragments are already methylated in plant and animal cells. Thus, the replicative DNA methylation in eukaryotes was discovered, and it was suggested that DNA methyltransferase may be a constituent of the DNA replicative complex.

DNA METHYLTRANSFERASES

The first mammalian DNA methyltransferases cloned were the mouse maintenance enzyme Dnmt1 (Bestor et al., 1988) and its closest human homologue DNMT1 (Yen et al., 1992). Interestingly, *DNMT1* mRNA was found to be most highly expressed in the brain. This high expression was rather surprising, considering the low proliferative potential of brain cells. A targeted mutation of *Dnmt1* does not affect morphology and growth rates of mouse embryonic stem cells (ESCs) in tissue culture, but in mouse embryos it results in severe developmental abnormalities and lethality (Li et al., 1992). The next mammalian *Dnmt* gene cloned was logically termed *Dnmt2* (Okano, Xie, & Li, 1998a). It contained all conserved methyltransferase motifs and, thus, could likely encode a functional cytosine methyltransferase. Dnmt2 has weak DNA methylation activity and seemed to be a dual-function protein capable of methylating both DNA and a cytosine residue in the anticodon loop of tRNA$_{Asp}$ (Jeltsch, Nellen, & Lyko, 2006). The next mouse genes cloned were *Dnmt3a* and *Dnmt3b*, encoding two highly similar proteins of 908 and 859 amino acids, respectively (Okano et al., 1998b). Cloned Dnmt3 proteins are the long-sought de novo DNA methyltransferases. *Dnmt3a* and *Dnmt3b* genes are highly expressed in ESCs and at a much lower levels in adult somatic tissues. Both were found to be required for genome-wide de novo methylation and essential for mammalian development (Okano, Bell, Haber, & Li, 1999).

The *Dnmt1* gene is highly expressed throughout the entire neuraxis at embryonic day 13 (E13) (Goto et al., 1994). Its expression in the brain decreases at E18 and postnatal day 1 (P1), although relatively high levels are retained in the forebrain. Expression is further attenuated at P7 throughout the entire brain, except for the granular layers of the cerebellum and the neuronal layer of the olfactory bulb. Only weak expression is discernible throughout the entire gray matter of the brain at P21 and later. Thus, the *Dnmt1* gene is expressed in postmitotic neurons, which after migration from the neurogenic zones have already reached their final destination in the brain. Its expression is maintained in most differentiated neurons in the adult and even aged brain, including the cerebellar granular layer and hippocampal neuronal layers. Considering that differentiated neurons are thought to no longer synthesize DNA, the persistent expression of *Dnmt1* indicates the occurrence of DNA methylation in neuronal cells in the absence of cell proliferation. After a conditional postmitotic neuron-specific knockout (KO) *Dnmt1* mutation in mice *Dnmt1*-deficient neurons were present in adult mutant brain at a constant percentage at all ages (Fan et al., 2001). When *Dnmt1* gene deletion occurred in ~30% neurons at midgestation, newborn mutants were born alive and normal, but the percentage of mutant cells in brain tissue was decreased significantly by P14 and not detectable at all by 3 weeks of age. Thus, Dnmt1 is required for the survival of mitotic neuronal precursor cells and their daughter cells, but not for the survival of postmitotic neurons.

Dnmt3a and Dnmt3b mRNAs are readily detectable in the newborn mouse cortex (Feng, Chang, Li, & Fan, 2005). In the cortex of adult mice *Dnmt3a* is expressed at a high level, whereas expression of *Dnmt3b* is not detectable. Thus, Dnmt3a and Dnmt3b play distinct roles in brain development. In the adult brain Dnmt3a is present in postmitotic neurons and oligodendrocytes and nearly absent in astroglial cells. A double knockout (DKO) of *Dnmt1* and *Dnmt3a* at P14 with the efficiency ~50% did not affect mouse life span and behavior, but significantly attenuated long-term potentiation in hippocampus and impaired learning and memory ability (Feng et al., 2010). Thus, Dnmt1 and Dnmt3a are required for normal synaptic plasticity. A gene expression analysis showed 84 genes to be upregulated and seven genes downregulated in DKO mice. An upregulated gene *Stat1* undergoes demethylation at the −895 to −1010 bp position of its promoter after deletion of *Dnmt1* and *Dnmt3a*. The cortex and hippocampus of DKO mice exhibit an ~20% reduction of total 5mC, suggesting that demethylation in DKO neurons could be widespread. Thus, Dnmt1 and Dnmt3a are essential for maintaining proper DNA methylation patterns at certain genomic loci in postmitotic CNS neurons.

Analysis of the *Dnmt3a* KO mice showed that Dnmt3a is required for neurogenesis (Wu et al., 2010). Two groups of Dnmt3a binding sites were found in the postnatal neuronal stem cell genome. The first group consists of genes in which CpG islands along their promotes are enriched with H3K4me3 (eg, *Dlx2*, *Gbx2*). On these genes Dnmt3a is generally excluded from H3K4me3-high, CpG-rich proximal promoter sequences, but it is enriched in flanking regions. The other group includes genes with CpG-poor promoters and low levels of H3K4me3 (eg, *Gfap*). The Dnmt3a binding sites on these genes frequently overlap with proximal promoter sequences. Thus, Dnmt3a may occupy and methylate defined genomic regions associated with both transcriptionally active and inactive genes. Gene ontology analysis showed that genes functionally related to neurogenesis are significantly enriched within H3K4me3-high group, whereas genes involved in development of nonneuronal lineage are enriched in H3K4me3-low group. Gene expression profiling of *Dnmt3a* KO mice identified both upregulated and down-regulated genes. The Dnmt3a binding sites within regulatory regions were detected in 942 of these genes; notably, genes with known roles in postnatal neurogenesis (eg, *Dlx2*, *Sp8*, and *Neurog2*) were downregulated, whereas those involved in astroglial and oligo-dendroglial differentiation (eg, *Sparcl1*, *Nkx2-2*) were upregulated. Thus, Dnmt3a bind-ing can either promote transcription (neurogenic genes) or repress it (glial differentiation genes). Dnmt3a deficiency was found to increase the H3K27me3 levels associated with many downregulated targets. Antagonism between Dnmt3a and H3K27me3 may facili-tate the expression of many H3K4me3-high targets by reducing H3K27me3 levels. Apparently, Dnmt3a not only mediates gene repression by methylating proximal pro-moters, but also promotes transcription of neurogenic genes by antagonizing Polycomb repression through nonproximal promoter methylation.

DNA DEMETHYLATION

Twenty-five years ago it was shown in vitro, that a methylated rat α-actin gene promoter transfected in L8 myoblasts is demethylated in the absence of DNA replication (Paroush et al., 1990). This process occurred in two steps: within a few hours on one DNA strand and after a ≥48-h delay on the complementary strand. Genetic analysis revealed the existence of *cis*-acting elements required for demethylation. The recognition of such sites early in the cell differentiation process probably leads to the demethylation required to activate gene transcription.

A transient DNA demethylation in Friend erythroleukemia cells induced to differ-entiation was accompanied by incorporation of deoxy[5-³H]cytidine, but not of deoxy[6-³H]adenosine, into preexistent DNA chains (Razin et al., 1986). Thus, demeth-ylation of DNA seemed to be achieved by an enzymatic mechanism whereby 5mC was replaced by cytosine.

Incubation of hemimethylated oligonucleotide DNA substrates with nuclear extracts of chicken embryo promoted active demethylation of 5mCpGs by a nucleotide excision repair mechanism (Jost, 1993). The first step of demethylation was the formation of spe-cific 5′ nick at the 5mC residue. Nicks also occurred on symmetrically methylated CpGs, but they resulted in breakage of the oligonucleotides with no repair. Nicks were strictly 5mCpG specific and did not occur on CpG, 5mCpC, 5mCpT, 5mCpA, or 6mApT. Purification of the demethylation activity revealed it to be a combination of 5mC-DNA glycosylase and apyrimidinic endonuclease (Jost, Siegmann, Sun, & Leung, 1995). The purified 5mC-DNA glycosylase also possessed a mismatch-specific thymine-DNA glycosylase (TDG) activity. It had a ≥6-fold preference for hemimethylated DNA substrates over symmetrically methylated substrates. Activity of the purified 5mC-DNA glycosylase was abolished by treatment of either proteinase K or ribonuclease A, suggesting it to belong to a protein- and RNA-enriched complex (Frémont et al., 1997). Indeed, RNA molecules, necessary for enzymatic activity, were found in purified 5mC-DNA glycosylase. Their cloning and sequencing revealed variable sequences of 200–600 nt unrelated to known RNAs or to each other, potentially reflecting interactions with long noncoding RNAs (Jost, Frémont, Siegmann, & Hofsteenge, 1997). The common feature of all clones analyzed was a high CpG density. On average, they have one CpG per each 14 nt and a CpG/GpC ratio of 1.1. It was shown that at least 4 nt sequences complementary to target methylated sites, mCpG and two adjacent nucleotides at each side, are required for efficient targeting in the demethylation reaction. Of the 16 possible NpCpGpN sequences, between 75% and 100% were present in each. The longest clone (618 nt) contained all of them and thus could serve as a universal targeting sequence. The different RNAs tightly linked to 5mC-DNA glycosylase were then suggested to represent transcripts from CpG islands, which should remain unmethylated (Jost et al., 1997).

A methylated DNA binding protein MBD2b, produced by in vitro translation of cloned cDNA, was found to transform 5mC residues in labeled substrate to C residues (Bhattacharya, Ramchandani, Cervoni, & Szyf, 1999). When demethylated DNA was subjected to CpG methyltransferase M.SssI, it was completely remethylated. The demethylase seemed to transform methylated cytosines in DNA to cytosines without disrupting the integrity of DNA chains. As the cleavage of a carbon–carbon bond requires high energy, direct demethylation of 5mC has been widely believed to be thermodynamically very unfavorable. In addition, formaldehyde has been shown to be released in demethylation reaction, indicating that the loss of the methyl group occurs due to oxidative demethylation of 5mC via the 5-hydroxymethylcytosine (5hmC) intermediate (Hamm et al., 2008). Unfortunately, direct demethylation of DNA by MBD2 could not be reproduced by others, who found MBD2 to participate in transcription repression through the recruitment of histone deacetylase (HDAC) to MeCP1 complex in HeLa cells (Hendrich, Guy, Ramsahoye, Wilson, & Bird, 2001; Ng et al., 1999).

Three proteins of the Gadd45 family are widely known to be involved in numerous biological processes, such as DNA damage responses, cell cycle control, senescence, apoptosis, and nucleotide excision repair (Salvador, Brown-Clay, & Fornace, 2013). A screen of *Xenopus* expression cDNA library for sequences able to reactivate the transcription of a methylation-silenced luciferase reporter gene resulted in isolation of Gadd45a (Barreto et al., 2007). Its reactivating activity seemed to be sequence and cell type independent. In human embryonic kidney 293 cells *Gadd45a* transfection led to a reduction of 5mC from 2.1% to 0.9% both in dividing and serum-starved nonproliferating cells, demonstrating active demethylation. An endonuclease XPG responsible for the 3′ incision during nucleotide excision repair and a DNA helicase were required for Gadd45a-mediated DNA demethylation. Thus, Gadd45a acted by promoting repair of methylated DNA sequences. In zebrafish embryos 5mC removal in vivo proceeds via coupled activities of activation-induced deaminase (AID) that converts 5mC to T and of Mbd4, a G:T mismatch-specific TDG containing a methyl-CG-binding domain, whereas Gadd45 serves as a nonenzymatic supporting factor (Rai et al., 2008). In mice, TDG KO is embryonically lethal (Cortellino et al., 2011), and TDG is involved in protection of CpG islands from hypermethylation and in active demethylation of tissue-specific developmentally and hormonally regulated promoters. In these roles TDG interacts with AID and Gadd45a.

A new twist in the DNA demethylation saga has begun with the discovery of the modified DNA base 5hmC in animal DNA (Kriaucionis & Heintz, 2009; Tahiliani et al., 2009). 5hmC is enriched exclusively in the brain, with higher abundance in the cortex and brainstem (Kriaucionis & Heintz, 2009). Three 2-oxoglutarate– and Fe(II)-dependent oxygenase human enzymes, TET1, TET2, and TET3, and their homologues in other animals, produce 5hmC through hydroxylation of the methyl group of 5mC (Tahiliani et al., 2009). Some 5mCs in mammalian cells are oxidized by TET proteins to 5hmC, which can be either deaminated to 5-hydroxymethyluracil (5hmU) by AID/APOBEC deaminases or further oxidized to 5-formylcytosine (5fC) and then to 5-carboxylcytosine (5caC). The 5hmU and 5caC are removed by the DNA glycosylase TDG, and the gap is refilled by unmethylated C through the BER pathway (Gong & Zhu, 2011; He et al., 2011). Also, Gadd45a has been demonstrated to directly interact with TDG and stimulate the removal of 5fC and 5caC from DNA (Li et al., 2015). KO of both *Gadd45a* and *Gadd45b* in mouse ESCs led to hypermethylation of genomic loci, most of which are targets for TDG and show 5fC enrichment in TDG-deficient cells. Thus, the DNA demethylation effects of Gadd45a could be mediated by TDG activity. These findings illustrate the increasing complexities of active DNA demethylation pathways in animals and, particularly, within the brain.

SPECIFICITY OF DNA METHYLATION

For quite a long time cytosine methylation in DNA was believed to occur mainly, if not exclusively, at CpG dinucleotides (Doskočil & Šorm, 1962; Grippo et al., 1968; Gruenbaum, Stein, Cedar, & Razin, 1981). Even before the methods of DNA sequencing were elaborated, we analyzed 5mC content in pyrimidine sequences isolated from DNA of various animals and plants by chemical hydrolysis and found that most 5mCs in animal DNAs are localized in the monopyrimidine fraction (Pu-5mC-Pu), whereas in plant DNAs significant 5mC quantities are present in sequences such as Pu-5mC-Pu, Pu-5mC-T-Pu, Pu-5mC-C-Pu, and Pu-5mC-5mC-Pu (Kirnos et al., 1981). Thus, animal DNAs seemed to be methylated mostly at CpG dinucleotides, whereas methylation of plant DNAs occurred both at CpG and CpHpG sites. According to our data, in plants up to ~30% 5mCs were localized in non–CpG sequences. This estimation was in a good accordance with the data obtained at the same time by a nearest neighbor method

(Gruenbaum, Naveh-Many, et al., 1981). The presence of 5mC in non–CpG sites in animal DNA was a matter of confusion for quite a long time. In a nearest neighbor analysis of DNA methylation in human spleen DNA, more than 50% total 5mC was found in non–CpG dinucleotides (Woodcock, Crowther, & Diver, 1987). At the time, these data were ascribed to inherent artifacts of the nearest neighbor method used. The matter was resolved when 5mC detection by genomic sequencing techniques arrived. It was shown that in animal cells DNA could be methylated at CpHpG sequences, both de novo and by maintenance type activity, although non–CpG methylation still seemed to be a rather rare phenomenon compared with CpG methylation (Clark et al., 1995; Toth et al., 1990). With the advent of genome-wide approaches to DNA methylation analysis, the non–CpG methylation was shown to be most prevalent in ESCs compared with somatic tissue cells (Ramsahoye et al., 2000). The dominant form of non–CpG methylation in the pluripotent cell types is 5mCpA, whereas in somatic cells CpA, CpT, and CpC are very rare and about equally methylated (Ziller et al., 2011). The enzymes responsible for non–CpG methylation are Dnmt3a and Dnmt3b: KO of their encoding genes leads to a global reduction in non–CpG methylation. The methylation levels of CpG and non–CpG sites are quite distinct. CpG sites are either not methylated at all or fully methylated. In contrast, non–CpG sites have intermediate methylation levels with a median between 30% and 50% (Ziller et al., 2011). Surprisingly, considerable levels of non–CpG methylation were found in brains of mice (Xie et al., 2012) and humans (Varley et al., 2013). In mouse frontal cortex non–CpG and CpG methylations have different genomic distributions; hence, non–CpG methylation could not be a trivial byproduct of CpG methylation. CpHpG and CpHpH methylations negatively correlate with gene expression in the mouse frontal cortex, irrespective of their location in the promoter regions or gene bodies (Xie et al., 2012). Unlike ESCs, CpHpHs are more likely to be methylated than CpHpGs in the mouse and human brains (Varley et al., 2013; Xie et al., 2012). CpH methylation accumulates during early postnatal development to maximal levels of 1.3–1.5% at the end of adolescence before diminishing slightly during aging (Lister et al., 2013). The most rapid increase in DNA methylation occurs during primary phase of synaptogenesis (from 2 to 4 weeks in mouse and at the first 2 years in humans), followed by slower accumulation during later adolescence. In mice the accumulation of 5mCpH from 1 to 4 weeks coincides with a transient increase in expression of the *Dnmt3a* gene. In genomic DNA purified from a relatively homogeneous population of granule neurons of the adult mouse dentate gyrus, ~25% of all 5mCs were found in mCpH sites, ~4% in mCpHpGs, and ~21% in mCpHpHs (Guo et al., 2014). A majority (~83%) of the human CpH-methylated genes had orthologs that were also CpH methylated in the mouse brain. Thus, the brain CpH methylation marks the conserved sets of genes in both mice and humans. The motif analysis identified a prominent CpApC preference for CpH methylation in neurons, predicting asymmetric methylation patterns on two DNA strands. A periodicity of ~180 bp for neuronal mCpHs was observed, suggesting a relationship to nucleosome positioning. Notably, mCpHs preferentially reside in regions of low CpG density. Both neuronal CpG and CpH methylation are inversely correlated with gene expression throughout the 5′ upstream, gene-body, and 3′ downstream regions. Thus, CpH methylation is a new layer of epigenetic modulation of the neuronal genome. This concept is reviewed in further detail by Mukamel and Lister in Chapter 4.

FUNCTIONAL ROLES OF DNA METHYLATION

Species, tissue, and age specificity

We happened to be among the first, who in a rather skeptical general attitude, asked the question, "What is the biological significance of DNA methylation?". Therefore, we endeavored to find various illustrative and adequate biological models that could be used for getting evidence of the significance of this genome modification. We learned a long time ago that DNA methylation is species specific. In many invertebrates the methylation degree of the genome is very low; as we have already mentioned, 5mC amount in *Drosophila* DNA is so small that it could not be detected for a very long time, whereas in vertebrate DNAs it is always present in noticeable amounts (Vanyushin, Tkacheva, et al., 1970); in plant DNA this base cannot even be called a minor base because its content is quite comparable with cytosine content (Vanyushin & Belozersky, 1959).

We have established that along with species specificity there are also tissue and age specificities of DNA methylation in animals (Berdyshev et al., 1967; Vanyushin, Mazin, et al., 1973; Vanyushin, Nemirovsky, et al., 1973; Vanyushin, Tkacheva, et al., 1970). These findings allowed us first to declare that DNA methylation should be a mechanism of regulation of gene expression and cell differentiation (Vanyushin, Mazin, et al., 1973). We also found that DNA methylation in mitochondria and the nucleus of one and the same animal or plant cell is different (Vanyushin, Alexandrushkina, & Kirnos, 1988; Vanyushin & Kirnos, 1974). Thus, the subcellular (organelle) specificity of DNA methylation was discovered. These data have drawn an attention of many investigators and pushed ahead the intense study of DNA methylation worldwide. Significant reductions in the genomic 5mC content upon aging were observed in mouse tissues (Wilson, Smith, Mag, & Cutler, 1987). The 5mC levels in DNA of different tissues (eg, brain, liver, intestinal mucosa) were observed to correlate with the chronological age. Now the age-dependent changes in DNA methylation are quite obvious, and some investigators are inclined even to consider DNA methylation degree as a sort of biological clock that measures the chronological age and forecasts life span (Florath, Butterbach, Müller, Bewerunge-Hudler, & Brenner, 2014). Distortions in DNA methylation may lead to premature aging.

Gene expression

Waalwijk and Flavell (1978) were first to use the isoschizomeric restriction endonuclease pair *Hpa*II and *Msp*I to test for the presence of 5mC in individual gene sequence. Both enzymes recognize the same site CCGG, but *Hpa*II does not cleave the sequence methylated at the internal C, whereas *Msp*I is not affected by such methylation. In most somatic tissues, including erythroid and nonerythroid tissues, a CCGG site present in the major intron of the rabbit β-globin gene seemed to be ~50% methylated. The same site seemed to be 100% methylated in sperm DNA and ~80% methylated in brain DNA. It was the first direct indication on reverse correlation between DNA methylation and gene activity. The next confirmation came from the Walter Doerfler's laboratory, where an inverse correlation between the extent of methylation and transcription of integrated adenovirus 12 genes in transformed hamster cells was noted (Sutter & Doerfler, 1980). We found that the patterns of DNA methylation in the rat liver change significantly upon induction of gene expression with hydrocortisone (Romanov & Vanyushin, 1981; Vanyushin, Nemirovsky, et al., 1973). These initial studies elicited an avalanche of similar investigations on a large number of eukaryotic genes. Now, it is widely believed that specific promoter methylation plays a crucial role in the stable silencing of eukaryotic genes, although the function of DNA methylation seems to vary with genomic sequence context, and the relationship between DNA methylation and transcription is more nuanced than it was realized at first (Jones, 2012). The connection between promoter methylation and inhibition of its activity is not obligatory. In most cases DNA methylation of promoter region does interfere with transcription by inhibiting binding of respective protein factors. However, there are some examples of positive effects of promoter methylation on transcription (Gustems et al., 2014). DNA methylation can either inhibit or stimulate binding of transcription regulatory proteins, and the effects of the binding could be either activation or repression of gene activity, dependent on protein function.

Learning and memory

The first indication that DNA methylation may be involved in brain activity came from our laboratory. We found that rat learning variably affects the neuronal DNA methylation levels in different regions of the rat brain (Guskova et al., 1977; Vanyushin et al., 1974, 1977). Two models of learning were used. In the first model the male rats were trained to a simple conditioned movement reflex to a light stimulus. Then DNA samples were isolated from neocortex, hippocampus, and cerebellum, and 5mC levels were analyzed by a thin-layer chromatography method. The 5mC level in the cerebellum DNA was not affected (0.99 ± 0.01 mol % in conditioned animals vs 0.96 ± 0.06 mol % in the control animals). In the neocortex DNA the 5mC level was increased by learning from 1.07 ± 0.07 to 1.45 ± 0.13 mol %. Similarly, in hippocampus the 5mC level was increased from 1.15 ± 0.01 to 1.83 ± 0.04 mol %. Thus, DNA methylation levels were variable in different brain divisions and were differently affected by learning. Specifically, learning did affect the genome methylation in brain divisions known to be involved in memory formation and maintenance. These effects were most pronounced at the early steps of conditioning, whereas at later steps there was only a small increase in 5mC levels. In the second model of learning the rats were trained to a Pavlovian food-conditioning reflex. At the early stages (20 and 50 min) of learning the methylation levels of the neocortex

DNA were increased by 22% and 26%, respectively; those of the hippocampus DNA by 18% and 7%,respectively; and those of the cerebellum DNA by 17% and 15%, respectively. At 24h the neocortex and hippocampus DNAs were still hypermethylated by ~10%, whereas the cerebellum DNA was not hypermethylated at all. In the active control animals that received equivalent numbers of uncombined stimuli, small hypermethylation was observed in the neocortex and cerebellum DNAs after 20 and 50min, whereas the hippocampus DNA was not hypermethylated. By 24h the methylation levels of the neocortex and cerebellum DNAs return to control values. Thus, functional activity of brain structures seemed to be accompanied by reversible changes in DNA methylation. Whether these changes represent a mechanism of the gene expression modulation was unknown at the time, although such suggestion seemed reasonable. Regardless, these data allowed us to first announce that brain (neuronal) DNA is involved in the memory formation (Vanyshin et al., 1974). An investigation of DNA synthesis in the rat brain on learning was then conducted to elucidate the nature of DNA hypermethylated on learning and the mechanisms of its subsequent demethylation (Ashapkin et al., 1983). The rat conditioning training, in both food-reward and passive defense reflex models, was found to be accompanied by an increase in DNA synthesis intensity (incorporation of [^{3}H]thymidine) in the brain cortex, whereas no effect was observed in cerebellum. The DNA synthesis induced was found to be selective in respect to various genome sequences and thus could not be a part of DNA replication. We have suggested it to represent an activated DNA repair synthesis involved in local DNA demethylation and chromatin remodeling. This could explain the observed reversibility of DNA methylation.

For a long time the topic of DNA methylation in relation to neurological memory was abandoned. It reappeared many years later, when the nature of genes involved in memory formation and maintenance was elucidated. The activity-dependent transcription of plasticity-related gene *Bdnf* was shown to be promoted by active demethylation at particular CpG sites in its promoter in postmitotic neurons from mouse cortex (Martinowich et al., 2003). The KCl treatment of cultured neurons led to membrane depolarization, activation of *Bdnf* transcription, and a decrease in methylation of CpG sites in its promoter. Elevated *Bdnf* expression in the *Dnmt1$^{-/-}$* mutant mouse brains was correlated with almost complete demethylation of these CpG sites. A partial dissociation of MeCP2 and more tight association of CREB were shown to occur and may account for induced *Bdnf* expression. In addition, less H3K9me2 and more H3K4me2 and acetylated H3 and H4 histones were associated with *Bdnf* promoter after depolarization. Thus, DNA methylation and chromatin remodeling seemed to play critical roles in regulating gene transcription in response to neuronal activity. Treatment of the hippocampal slices with the Dnmt inhibitor zebularine resulted in demethylation of specific CpG island sequences in promoters of two genes involved in memory, *reelin* and *Bdnf* (Levenson et al., 2006). Activation of protein kinase C (PKC) in the hippocampal slices with phorbol-12,13-diacetate led to a rapid demethylation of the same sequence in the *reelin* promoter and an increase in the *Dnmt3a* gene expression in Area CA1 of the hippocampus. The expression of immediate early gene *c-fos* was also significantly increased. These results showed that the expression of *Dnmt3a* in the hippocampus is regulated by PKC signaling cascade and probably plays a role in synaptic plasticity.

Increased levels of *Dnmt3a* and *Dnmt3b* mRNAs were found in the hippocampal area CA1 of rats trained to contextual fear conditioning relative to control animals exposed only to the novel context of the experimental chamber (Miller & Sweatt, 2007). Dnmt inhibitors, 5-aza-dC and zebularine, injected directly into area CA1 immediately after training significantly impaired memory retention 24 h later. Thus, the hippocampal Dnmt activity seems to be necessary for memory consolidation. The methylation level of the protein phosphatase 1 (PP1) gene, known to suppress learning and memory, was greatly increased at 1 h after training. At the same time, the methylation levels of *reelin* gene, which promotes synaptic plasticity and memory, were significantly decreased and its transcription increased. Hence, some DNA demethylation mechanism acting in an activity-dependent manner during memory consolidation must be present in hippocampal neurons. The methylation levels of both *reelin* and *PP1* returned to control levels within a day of training. Thus, DNA methylation changes after training are both rapid and reversible. Very similar findings were reported for *Bdnf* (Lubin et al., 2008). *Bdnf* mRNA levels in the area CA1 of rat hippocampus were increased within 30 min of fear conditioning, still more increased at 2 h, and returned to baseline levels at 24 h. The exon-specific *Bdnf* mRNA levels were correlated with decreased methylation of respective CpG islands. Thus, DNA methylation controls the exon-specific readout of the *Bdnf* gene. A surprising complexity in the control of DNA methylation at the *Bdnf* gene locus, involving decreases and increases in methylation at individual transcription initiation sites, was noted. DNA methylation in the adult hippocampus seemed to play a role in memory consolidation, but not in long-term memory storage.

In a rat model of contextual fear conditioning the immediate early gene *Egr1* was found to be demethylated in adult neurons of the neocortex both in trained and active control animals at 1 h, 1 day, and 7 days after training (Miller et al., 2010). *Reelin* gene was hypermethylated in trained rats in 1 h after training. The hypermethylation was reduced at the later time points. The memory suppressor gene *calcineurin* (*CaN*) was not affected shortly after training, but highly hypermethylated at 1 and 7 days. Methylated CpGs were randomly distributed across the analyzed 0.5-kb segment of *CaN* promoter associated CpG island. An *N*-methyl-D-aspartate (NMDA) receptor antagonist MK-801 interfered with both acquisition of fear memory and hypermethylation of *CaN* and *reelin* at 7 days, without affecting *Egr1* methylation. Thus, *CaN* and *reelin* hypermethylations seemed to be a specific response to associative environmental signals. An infusion of the NMDA receptor antagonist D-(−)-2-amino-5-phosphonovaleric acid directly into dorsal hippocampus (area CA1) immediately before training interfered with both learning and

CaN and *reelin* methylations in the dorsomedial prefrontal cortex 7 days after training, indicating that a single hippocampus-dependent learning experience is sufficient to drive lasting, gene-specific methylation changes in the cortex. The fear memory and the methylation status of these genes were persistent at 30 days after training. Rats received intracortical infusions of Dnmt inhibitors at 30 days after training failed to display normal memory. Thus, DNA methylation in the dorsomedial prefrontal cortex is critical for remote memory stability. The same infusions 24 h after training, before the memory became reliant on the dorsomedial prefrontal cortex, did not interfere with fear memory at 2 and 30 days posttraining. Thus, cortical DNA methylation is triggered by a learning experience and is a perpetuating signal used by the brain to promote and maintain remote memories.

As stated previously, proteins of Gadd45 family play a role in the locus-specific DNA demethylation events. An electroconvulsive treatment (ECT) of mature neurons in mice dentate gyrus induced expression of *Gadd45b* gene (Ma et al., 2009). Spatial exploration of a novel environment, known to activate immediate early genes, also led to significant induction of *Gadd45b*. Most *Gadd45b*-positive cells also expressed *Arc*, a classic activity-induced immediate early gene. In vivo injection of an NMDA receptor antagonist abolished the induced *Gadd45b* and *Arc* expression. Adult *Gadd45b* KO mice seemed anatomically normal and exhibited identical NMDA receptor-dependent induction of immediate early genes at 1 h after ECT. The basal densities of proliferating cells in the dentate gyrus were similar in the wild-type (WT) and KO mice, but after ECT there was a 140% increase in the density of such cells in WT mice and only a 40% increase in KO littermates. Reducing expression of endogenous *Gadd45b* by a small hairpin RNA largely abolished induced, but not the basal proliferation of adult neural progenitors. Thus, *Gadd45b* plays an essential role in activity-induced, but not basal, proliferation of neural progenitors in the adult dentate gyrus. ECT markedly increased the total dendritic length and complexity of postmitotic neurons in dentate gyrus of adult mice. This ECT-induced dendritic growth was significantly attenuated in KO mice, whereas the basal level of dendritic growth was unaffected. Thus, *Gadd45b* is also essential for activity-induced dendritic development of newborn neurons in the adult brain. No significant global DNA demethylation in mature neurons of adult dentate tissue was detected after ECT in vivo. However, significant demethylation was found at specific regulatory regions of *Bdnf* and *Fgf1* genes. The basal levels of DNA methylation within these regions were similar in WT and KO mice, whereas ECT-induced DNA demethylation of these regions was almost completely abolished in KO mice. Thus, GADD45b is essential for activity-dependent demethylation and late-onset expression of specific secreted factors in the adult dentate gyrus. *Gadd45b* expression in hippocampus was upregulated during memory consolidation after contextual learning and associative fear training (Leach et al., 2012; Sultan, Wang, Tront, Liebermann, & Sweatt, 2012). *Gadd45b* KO mice had no significant changes in baseline behavior, but performed significantly lower compared with WT mice in the long-term memory test (24 h after contextual learning conditioning), but did not differ in short-term memory test (1 h after conditioning) (Leach et al., 2012). Somewhat contradictory results were obtained using a rotarod motor learning model: 24 h after training *Gadd45b*$^{-/-}$ mice on the hybrid background demonstrated significantly enhanced performance versus WT, implicating *Gadd45b* in motor memory consolidation, but not initial acquisition (Sultan et al., 2012). In cue-plus-context fear conditioning a significantly enhanced performance of KO mice compared with WT mice was also found at 24 h with mild and moderate, but not robust training. Whatever the reasons for these differences, both studies support the general view that *Gadd45b* is transcriptionally regulated by experience and also regulates memory capacity. Recently, an excision repair mechanism of DNA demethylation by Gadd45 proteins was demonstrated (Li et al., 2015). Thus, our earlier suggestion that region-specific DNA demethylation at learning can be mediated through DNA repair–like mechanisms (Ashapkin et al., 1983) gains an experimental support.

DNA methylome changes in the dentate gyrus granule neurons in the adult mouse hippocampus in vivo after synchronous neuronal activation by ECS were analyzed at single-nucleotide resolution by using a next-generation sequencing–based method (Guo et al., 2011). Of ~200,000 CpGs analyzed, 1892 and 1158 exhibited activity-induced de novo methylation and demethylation, respectively, at 4 h after ECS. Thus, methylation of at least ~1.4% CpGs is rapidly modifiable by neuronal activity in the adult brain. Some methylation changes were reversed by 24 h after ECS, but 31% activity-modified CpGs remained at their modified states. Pretreatment of animals with a highly selective NMDA receptor antagonist 3-(2-carboxypiperazin-4-yl)propyl-1-phosphonic acid abolished ECS-induced changes, confirming these modifications to be neuronal activity dependent. Infusion of either 5-aza-dC or RG108, two Dnmt inhibitors with distinct mechanisms of action, abolished activity-induced de novo methylation with no obvious effect on demethylation. Dnmt3a, but not other Dnmts, was upregulated by ECS, suggesting its potential role in the neuronal activity-induced de novo methylation. In contrast, activity-induced demethylation was abolished in the *Gadd45b* KO mice, consistent with its role in DNA demethylation. Bisulfite sequencing of five representative regions revealed that activity-induced modifications are highly site specific. Similar ECS-induced CpG modifications were observed in directly fluorescence-activated cell sorting–purified NeuN$^+$ postmitotic neuronal nuclei from the dentate gyrus, indicating that the CpG modifications observed are predominantly neuronal and independent of DNA replication. To further determine whether a physiological paradigm of neuronal stimulation could induce DNA methylation changes, adult mice were subjected to a 3-day course of voluntary running, and highly similar changes were observed. Thus, widespread changes in CpG methylation occur in postmitotic neurons in vivo in response to both chemical and physiological neuronal activation. Analysis of genomic location of activity-modified CpGs revealed a striking exclusion of methylation changes in CpG-dense regions for both activity-induced de novo methylation and demethylation; significant resistance of both gene-associated and intergenic CpG islands to activity-induced methylation changes was observed. Thus, the main targets of activity-induced acute modifications are low-density CpGs. Activity-modified CpGs are underrepresented in the 5′ upstream regions (putative promoters), exons and 3′ downstream regions of the genes, but slightly enriched in introns. Intergenic CpGs (≥5 kb away from any known genes) are most susceptible to methylation changes induced by neuronal activity. Although activity-modified CpGs are enriched in intergenic regions, 1819 activity-modified CpGs were mapped to 1518 genes. The methylation changes located in their 5′ upstream regions (putative promoters) were modestly but significantly anticorrelated with changes in expression. In contrast, no significant correlation was detected between methylation changes in other gene parts and their expression. Thus, the activity-induced methylation changes may regulate gene expression in a highly context-dependent manner and may have other roles, besides transcription regulation. The 1518 genes associated with the activity-modified CpGs are significantly enriched in genes that are expressed in the brain. The activity-modified CpGs are also preferentially associated with the alternative splicing variants, suggesting a potential role of DNA methylation changes in regulating alternative splicing in neurons. A gene ontology analysis revealed significant overrepresentations of genes involved in synaptic function, protein phosphorylation, neuronal differentiation, and the calcium signaling pathway. Some of them are enriched also in activity-regulated genes at the mRNA level. Surprisingly, multiple genes encoding Notch signaling components exhibited CpG methylation and expression changes. Identification of the Notch signaling pathway as a novel epigenetic target of neuronal activity in mature neurons supports its emerging role in synaptic plasticity and long-term memory (Pierfelice, Alberi, & Gaiano, 2011). The role of DNA modifications in learning and memory is explored in further detail in Chapters 5 and 8.

Early life stress behavior

A striking example of epigenetic adaptive phenomenon came from the studies of rat maternal nursing behavior (licking and grooming and arched-back nursing) effects on the offspring responses to stress (Weaver et al., 2004). As adults, the offspring of high-nursing (HN) mothers appeared less fearful and show more modest hypothalamus–pituitary axis (HPA) responses to stress compared with the offspring of low-nursing (LN) mothers. Intriguingly, the biological offspring of LN mothers reared by HN females resemble the normal offspring of HN mothers and vice versa. The adult offspring of HN mothers show increased hippocampal glucocorticoid receptor gene *Nr3c1* expression and enhanced glucocorticoid feedback sensitivity on hypothalamic corticotropin-releasing hormone release and HPA response to stress compared with offspring of LN mothers. This increase in *Nr3c1* gene expression has been shown to be mediated by 5-hydroxytryptamine (serotonin) (5-HT) activity at 5-HT$_7$ receptors and subsequent activation of cAMP-dependent protein kinase activity. This effect is also accompanied by an increased hippocampal expression of nerve growth factor–inducible protein A (NGFI-A, a transcription factor also known as egr-1, krox-24, zenk, and zif-268). The noncoding exon 1 region of the hippocampal *Nr3c1*, exon 1_7, includes a promoter region containing a binding site for NGFI-A. Splice variants of the *Nr3c1* mRNA containing the exon 1_7 sequence are found predominantly in brain, and the expression of *Nr3c1* mRNAs containing the exon 1_7 sequence is increased in the offspring of HN mothers. In hippocampal DNA a CpG site within NGFI-A consensus sequence was found to be always methylated in the offspring of LN mothers and rarely methylated in the offspring of HN females. When biological offspring of HN and LN mothers was cross-fostered, the patterns of the exon 1_7 promoter methylation in offspring were associated with the rearing female, not the biological mother. Thus, variations in maternal care directly alter the methylation status of the *Nr3c1* gene in offspring brain DNA. Just before birth (E20), the entire sequence of the exon 1_7 *Nr3c1* promoter is unmethylated in both offspring groups. In one day after birth (P1), this region is de novo methylated to the same extent in both animal groups. The differences in the methylation status emerge between P1 and P6, the period when differences in the maternal behavior are apparent. By P6, the NGFI-A response element CpG dinucleotide is effectively demethylated in the HN group, but not in the LN group. This difference remains consistent to adulthood (P90). Thus, the group difference in DNA methylation occurs as a function of a maternal behavior over the first week of life. It was shown that there is a significantly greater histone H3K9ac association and threefold greater binding of NGFI-A protein to the hippocampal exon 1_7 *Nr3c1* promoter in the offspring of HN mothers compared with offspring of LN mothers. Central infusion of the HDAC inhibitor trichostatin A enhances histone H3K9 acetylation of the exon 1_7 *Nr3c1* promoter in the offspring of the LN mothers, increases NGFI-A binding to its cognate sequence, induces hypomethylation of CpGs in the promoter, and eliminates the maternal effect on hippocampal *Nr3c1* expression and the HPA response to stress.

Similar results were obtained in a study of the estrogen receptor gene *ERα* expression and methylation in medial preoptic area (MPOA) of the hypothalamus in the offspring of HN and LN mothers (Champagne et al., 2006). *ERα* expression in the MPOA of the P6 offspring of HN mothers was found to be significantly increased compared with that of the offspring of LN mothers, whereas the methylation levels of CpG sites across the *ERα* promoter were significantly higher in the offspring of LN mothers. By a cross-fostering paradigm a highly significant effect of rearing female was demonstrated. Among the CpG sites differentially methylated was one contained within a consensus Stat5b binding site. A significantly greater binding of Stat5b to the *ERα* promoter in the adult offspring of HN compared with LN mothers was demonstrated. Since *ERα* expression is regulated through activation of the Janus kinase–Stat5b pathway, inhibition of Stst5b binding to its methylated binding site could be a mechanism, by which DNA methylation suppresses $ER\alpha$ expression. Thus, maternal care is associated with increased $ER\alpha$ expression, together with demethylation of its promoter and enhanced Stat5b binding.

The experience of the mother can also be translated through an epigenetic inheritance into phenotypic variation in the offspring, resulting in the transmission of adaptive responses across generations. The effect of maternal care on the hippocampal transcriptome seems to be gene specific. Of total hippocampus transcriptome only 253 transcripts (0.81%) were found to be upregulated and 50 transcripts (0.16%) downregulated in the offspring of HN mothers compared with offspring of LN mothers, whereas 30,796 (99%) transcripts remained unaltered (Weaver, Meaney, & Szyf, 2006). Thus, maternal care during early life programs the expression of hundreds of genes in the adult offspring.

Familial childhood adversities are associated with altered HPA stress responses and increased risk for multiple forms of psychopathology in humans. There is evidence for decreased hippocampal glucocorticoid receptor expression at several psychopathological conditions associated with suicide, including schizophrenia and mood disorders. Suicide is also strongly associated with a history of the childhood abuse and neglect. An investigation of the glucocorticoid receptor gene NR3C1 expression in hippocampal samples obtained from suicide victims showed a significant reduction of NR3C1 mRNA in suicide victims with a history of childhood abuse relative to nonabused suicide victims or controls (McGowan et al., 2009). A significant effect on NR3C1 promoter methylation was also found.

Early prenatal stress (E-PS) in mice was found to significantly increase the corticotropin-releasing hormone (Crh) gene expression in the amygdala central nucleus and decrease Nr3c1 gene expression in the hippocampal CA3 area and dentate gyrus (Mueller & Bale, 2008). In the hypothalamus of E-PS males, reduced methylation of the Crh gene promoter and increased methylation of exon 1_7-specific part of the Nr3c1 gene promoter were observed. Thus, dysregulation of Crh and stress pathway programming may be a key contributor to the stress-sensitive phenotype in E-PS male mice.

Early life stress (ELS) during the first 10 days of life in mice led to sustained hyperactivity of the HPA. ELS induces increased expression of pituitary pro-opiomelanocortin (Pomc) mRNA. Pomc expression is induced by the hypothalamic neuropeptides arginine vasopressin (Avp) and Crh, and all of them are under negative feedback control by the glucocorticoid receptor. Conspicuously, levels of Nr3c1 mRNA in the hippocampus, hypothalamic paraventricular nucleus (PVN) and pituitary were either unchanged or upregulated in ELS-treated mice, arguing against impaired corticosterone feedback as the primary cause of the observed increases in Pomc expression and glucocorticoid secretion (Mutgatroyd et al., 2009). ELS did not influence the hypothalamic Crh mRNA expression, but resulted in a significant upregulation of Avp mRNA. The changes in Avp expression persisted for at least 1 year and were restricted to the parvocellular subpopulation of neurons in the PVN that drive the pituitary–adrenal axis. ELS also produced long-lasting memory deficits in an inhibitory avoidance task. Of four *Avp* gene CpG islands, promoter CGI1 and exon CGI2 are sparsely methylated in control mice, whereas intergenic region (includes a composite enhancer region between tail-to-tail–orientated *Avp* and *oxytocin* genes) CGI3 has high levels of CpG methylation, and CGI4 and the adjacent *oxytocin* tissue–specific enhancer region have a less-dense methylated pattern with few, irregularly spaced highly methylated CpG dinucleotides. A hypomethylation of multiple CpGs throughout the downstream *Avp* enhancer region was found in PVN tissue of adult ELS mice. Of the 11 CpGs significantly hypomethylated in 6-week-old ELS mice, 7 in CGI3 had methylation patterns that strongly correlated with *Avp* mRNA levels. This means that the ELS-induced changes in the methylation status of relevant CpG residues are persistent and sustain elevated *Avp* expression.

Thus, adverse events in early life can leave persistent epigenetic marks on specific genes that may prime susceptibility to neuroendocrine and behavioral dysfunctions. In general, postmitotic epigenetic modifications regulate gene expression and related neuronal function, which can serve to facilitate or disfavor physiological and behavioral adaptations.

CONCLUSION

DNA methylation, for a long time regarded as a stable DNA modification that controls cellular differentiation, is now widely accepted to be dynamically regulated in the nervous system and plays an essential role in memory formation and adaptation to an ever-changing environment. Dynamic changes in DNA methylation occur in postmitotic neurons; the methylation-mediated chromatin remodeling may play critical roles in gene expression modulation involved in long-lasting neuronal responses. DNA methylation represents one of the most permanent mechanisms of cellular memory. In support of this view, most differentiated cells have a rather small activity of DNMTs. Although the mature brain consists mainly of terminally differentiated postmitotic cells, surprisingly, high levels of DNA methylation and demethylation activities persist throughout the whole life.

ACKNOWLEDGMENT

This work was supported by Russian Science Foundation grant no. 14-50-00029.

REFERENCES

Ashapkin, V. V., Antoniv, T. T., & Vanyushin, B. F. (1995). Methylation-dependent binding of wheat nuclear proteins to promoter region of ribosomal RNA genes. *Gene, 157*, 273–277.
Ashapkin, V. V., Romanov, G. A., Tushmalova, N. A., & Vanyushin, B. F. (1983). Selective synthesis of DNA in the rat brain induced by learning. *Biokhimiya, 48*, 355–362.
Barreto, G., Schäfer, A., Marhold, J., Stach, D., Swaminathan, S. K., Handa, V., et al. (2007). Gadd45a promotes epigenetic gene activation by repair-mediated DNA demethylation. *Nature, 445*, 671–675.

Bashkite, E. A., Kirnos, M. D., Kiryanov, G. I., Aleksandrushkina, N. I., & Vanyushin, B. F. (1980). Replication and methylation of DNA in tobacco cell suspension culture and the influence of auxin. *Biokhimiya*, *45*, 1448–1456.

Berdyshev, G. D., Korotaev, G. K., Boyarskikh, G. V., & Vanyushin, B. F. (1967). Nucleotide composition of DNA and RNA from somatic tissues of humpback salmon and its changes during spawning. *Biokhimiya*, *32*, 988–993.

Bestor, T., Laudano, A., Mattaliano, R., & Ingram, V. (1988). Cloning and sequencing of a cDNA encoding DNA methyltransferase of mouse cells: the carboxyl-terminal domain of the mammalian enzymes is related to bacterial restriction methyltransferases. *Journal of Molecular Biology*, *203*, 971–983.

Bhattacharya, S. K., Ramchandani, S., Cervoni, N., & Szyf, M. (1999). A mammalian protein with specific demethylase activity for mCpG DNA. *Nature*, *397*, 579–583.

Champagne, F., Weaver, I., Diorio, J., Dymov, S., Szyf, M., & Meaney, M. J. (2006). Maternal care associated with methylation of the estrogen receptor-α1b promoter and estrogen receptor- α expression in the medial preoptic area of female offspring. *Endocrinology*, *147*, 2909–2915.

Clark, S. J., Harrison, J., & Frommer, M. (1995). CpNpG methylation in mammalian cells. *Nature Genetics*, *10*, 20–27.

Cortellino, S., Xu, J., Sannai, M., Moore, R., Caretti, E., Cigliano, A., et al. (2011). Thymine DNA glycosylase is essential for active DNA demethylation by linked deamination-base excision repair. *Cell*, *146*, 67–79.

Culp, L. A., Dore, E., & Brown, G. M. (1970). Methylated bases in DNA of animal origin. *Archives of Biochemistry and Biophysics*, *136*, 73–79.

Doskočil, J., & Šorm, F. (1962). Distribution of 5-methylcytosine in pyrimidine sequences of deoxyribonucleic acids. *Biochimica et Biophysica Acta*, *55*, 953–959.

Dunn, D. B., & Smith, J. D. (1955). Occurrence of a new base in the deoxyribonucleic acid of a strain of *Bacterium coli*. *Nature*, *175*, 336–337.

Fan, G., Beard, C., Chen, R. Z., Csankovszki, G., Sun, Y., Siniaia, M., et al. (2001). DNA hypomethylation perturbs the function and survival of CNS neurons in postnatal animals. *The Journal of Neuroscience: the Official Journal of the Society for Neuroscience*, *21*, 788–797.

Feng, J., Chang, H., Li, E., & Fan, G. (2005). Dynamic expression of de novo DNA methyltransferases Dnmt3a and Dnmt3b in the central nervous system. *Journal of Neuroscience Research*, *79*, 734–746.

Feng, J., Zhou, Y., Campbell, S. L., Le, T., Li, E., Sweatt, J. D., et al. (2010). Dnmt1 and Dnmt3a are required for the maintenance of DNA methylation and synaptic function in adult forebrain neurons. *Nature Neuroscience*, *13*, 423–430.

Florath, I., Butterbach, K., Müller, H., Bewerunge-Hudler, M., & Brenner, H. (2014). Cross-sectional and longitudinal changes in DNA methylation with age: an epigenome-wide analysis revealing over 60 novel age-associated CpG sites. *Human Molecular Genetics*, *23*, 1186–1201.

Frémont, M., Siegmann, M., Gaulis, S., Matthies, R., Hess, D., & Jost, J.-P. (1997). Demethylation of DNA by purified chick embryo 5-methylcytosine-DNA glycosylase requires both protein and RNA. *Nucleic Acids Research*, *25*, 2375–2380.

Gold, M., & Hurwitz, J. (1963). The enzymatic methylation of the nucleic acids. *Cold Spring Harbour Symposia of Quantitative Biology*, *28*, 149–156.

Gong, Z., & Zhu, J.-K. (2011). Active DNA demethylation by oxidation and repair. *Cell Research*, *21*, 1649–1651.

Goto, K., Numata, M., Komura, J.-I., Ono, T., Bestor, T. H., & Kondo, H. (1994). Expression of DNA methyltransferase gene in mature and immature neurons as well as proliferating cells in mice. *Differentiation*, *56*, 39–44.

Gowher, H., Leismann, O., & Jeltsch, A. (2000). DNA of *Drosophila melanogaster* contains 5-methylcytosine. *EMBO Journal*, *19*, 6918–6923.

Grippo, P., Iaccarino, M., Parisi, E., & Scarano, E. (1968). Methylation of DNA in developing sea urchin embryos. *Journal of Molecular Biology*, *36*, 195–208.

Gruenbaum, Y., Naveh-Many, T., Cedar, H., & Razin, A. (1981). Sequence specificity of methylation in higher plant DNA. *Nature*, *292*, 860–862.

Gruenbaum, Y., Stein, R., Cedar, H., & Razin, A. (1981). Methylation of CpG sequences in eukaryotic DNA. *FEBS Letters*, *124*, 67–71.

Guo, J. U., Ma, D. K., Mo, H., Ball, M. P., Jang, M.-H., Bonaguidi, M. A., et al. (2011). Neuronal activity modifies DNA methylation landscape in the adult brain. *Nature Neuroscience, 14*, 1345–1351.

Guo, J. U., Su, Y., Shin, J. H., Shin, J., Li, H., Xie, B., et al. (2014). Distribution, recognition and regulation of non-CpG methylation in the adult mammalian brain. *Nature Neuroscience, 17*, 215–222.

Guskova, L. V., Burtseva, N. N., Tushmalova, N. A., & Vaniushin, B. F. (1977). Level of methylated nuclear DNA in neurons and glia of rat cerebral cortex and its alteration upon elaboration of a conditioned reflex. *Doklady Akademii Nauk SSSR, 233*, 993–996.

Gustems, M., Woellmer, A., Rothbauer, U., Eck, S. H., Wieland, T., Lutter, D., et al. (2014). c-Jun/c-Fos heterodimers regulate cellular genes via a newly identified class of methylated DNA sequence motifs. *Nucleic Acids Research, 42*, 3059–3072.

Hamm, S., Just, G., Lacoste, N., Moitessier, N., Szyf, M., & Mamer, O. (2008). On the mechanism of demethylation of 5-methylcytosine in DNA. *Bioorganic & Medicinal Chemistry Letters, 18*, 1046–1049.

He, Y.-F., Li, B.-Z., Li, Z., Liu, P., Wang, Y., Tang, Q., et al. (2011). Tet-mediated formation of 5-carboxylcytosine and its excision by TDG in mammalian DNA. *Science, 333*, 1303–1307.

Hendrich, B., Guy, J., Ramsahoye, B., Wilson, V. A., & Bird, A. (2001). Closely related proteins MBD2 and MBD3 play distinctive but interacting roles in mouse development. *Genes & Development, 15*, 710–723.

Holliday, R., & Pugh, J. E. (1975). DNA modification mechanisms and gene activity during development. *Science, 187*, 226–232.

Hotchkiss, R. D. (1948). The quantitative separation of purines, pyrimidines and nucleosides by paper chromatography. *The Journal of Biological Chemistry, 175*, 315–332.

Jeltsch, A., Nellen, W., & Lyko, F. (2006). Two substrates are better than one: dual specificities for Dnmt2 methyltransferases. *Trends in Biochemical Sciences, 31*, 306–308.

Jones, P. A. (2012). Functions of DNA methylation: islands, start sites, gene bodies and beyond. *Nature Reviews Genetics, 13*, 484–492.

Jost, J.-P. (1993). Nuclear extracts of chicken embryos promote an active demethylation of DNA by excision repair of 5-methyldeoxycytidine. *Proceedings of the National Academy of Sciences of the United States of America, 90*, 4684–4688.

Jost, J.-P., Frémont, M., Siegmann, M., & Hofsteenge, J. (1997). The RNA moiety of chick embryo 5-methylcytosine-DNA glycosylase targets DNA demethylation. *Nucleic Acids Research, 25*, 4545–4550.

Jost, J.-P., Siegmann, M., Sun, L., & Leung, R. (1995). Mechanism of DNA demethylation in chicken embryos. *The Journal of Biological Chemistry, 370*, 9734–9739.

Kirnos, M. D., Aleksandrushkina, N. I., & Vanyushin, B. F. (1981). 5-Methylcytosine in pyrimidine sequences of plant and animal DNA: specificity of methylation. *Biokhimiya, 46*, 1458–1474.

Kiryanov, G. I., Kirnos, M. D., Demidkina, N. P., Alexandrushkina, N. I., & Vanyushin, B. F. (1980). Methylation of DNA in L cells on replication. *FEBS Letters, 112*, 225–228.

Kriaucionis, S., & Heintz, N. (2009). The nuclear DNA base 5-hydroxymethylcytosine is present in Purkinje neurons and the brain. *Science, 324*, 929–930.

Leach, P. T., Poplawski, S. G., Kenney, J. W., Hoffman, B., Liebermann, D. A., Abel, T., et al. (2012). Gadd45b knockout mice exhibit selective deficits in hippocampus-dependent long-term memory. *Learning & Memory, 19*, 319–324.

Levenson, J. M., Roth, T. L., Lubin, F. D., Miller, C. A., Huang, I.-C., Desai, P., et al. (2006). Evidence that DNA (cytosine-5) methyltransferase regulates synaptic plasticity in the hippocampus. *The Journal of Biological Chemistry, 281*, 15763–15773.

Li, E., Bestor, T. H., & Jaenisch, R. (1992). Targeted mutation of the DNA methyltransferase gene results in embryonic lethality. *Cell, 69*, 915–926.

Li, Z., Gu, T.-P., Weber, A. R., Shen, J.-Z., Li, B.-Z., Xie, Z.-G., et al. (2015). Gadd45a promotes DNA demethylation through TDG. *Nucleic Acids Research, 43*, 3986–3997.

Lister, R., Mukamel, E. A., Nery, J. R., Urich, M., Puddifoot, C. A., Johnson, N. D., et al. (2013). Global epigenomic reconfiguration during mammalian brain development. *Science, 341*, 1237905. http://dx.doi.org/10.1126/science.1237905.

Lubin, F. D., Roth, T. L., & Sweatt, J. D. (2008). Epigenetic regulation of *bdnf* gene transcription in the consolidation of fear memory. *The Journal of Neuroscience: the Official Journal of the Society for Neuroscience, 28*, 10576–10586.

Lyko, F., Ramsahoye, B. H., & Jaenisch, R. (2000). DNA methylation in *Drosophila melanogaster*. *Nature*, *408*, 538–540.

Ma, D. K., Jang, M.-H., Guo, J. U., Kitabatake, Y., Chang, M., Pow-anpongku, N., et al. (2009). Neuronal activity–induced Gadd45b promotes epigenetic DNA demethylation and adult neurogenesis. *Science*, *323*, 1074–1077.

Martinowich, K., Hattori, D., Wu, H., Fouse, S., He, F., Hu, Y., et al. (2003). DNA methylation–related chromatin remodeling in activity-dependent *Bdnf* gene regulation. *Science*, *302*, 890–893.

Mazin, A. L., & Vanyushin, B. F. (1988). Loss of CpG dinucleotides from DNA. 5. Traces of "fossil" methylation in *Drosophila* genome. *Molecular Biology (Moscow)*, *22*, 1399–1404.

McGowan, P. O., Sasaki, A., D'Alessio, A. C., Dymov, S., Labonte, B., Szyf, M., et al. (2009). Epigenetic regulation of the glucocorticoid receptor in human brain associates with childhood abuse. *Nature Neuroscience*, *12*, 342–348.

Miller, C. A., Gavin, C. F., White, J. A., Parrish, R. R., Honasoge, A., Yancey, C. R., et al. (2010). Cortical DNA methylation maintains remote memory. *Nature Neuroscience*, *13*, 664–666.

Miller, C. A., & Sweatt, J. D. (2007). Covalent modification of DNA regulates memory formation. *Neuron*, *53*, 857–869.

Mueller, B. R., & Bale, T. L. (2008). Sex-specific programming of offspring emotionality after stress early in pregnancy. *The Journal of Neuroscience: the Official Journal of the Society for Neuroscience*, *28*, 9055–9065.

Murgatroyd, C., Patchev, A. V., Wu, Y., Micale, V., Bockmühl, Y., Fischer, D., et al. (2009). Dynamic DNA methylation programs persistent adverse effects of early-life stress. *Nature Neuroscience*, *12*, 1559–1566.

Ng, H.-H., Zhang, Y., Hendrich, B., Johnson, C. A., Turner, B. M., Erdjument-Bromage, H., et al. (1999). MBD2 is a transcriptional repressor belonging to the MeCP1 histone deacetylase complex. *Nature Genetics*, *23*, 58–61.

Okano, M., Bell, D. W., Haber, D. A., & Li, E. (1999). DNA methyltransferases Dnmt3a and Dnmt3b are essential for de novo methylation and mammalian development. *Cell*, *99*, 247–257.

Okano, M., Xie, S., & Li, E. (1998a). Dnmt2 is not required for de novo and maintenance methylation of viral DNA in embryonic stem cells. *Nucleic Acids Research*, *26*, 2536–2540.

Okano, M., Xie, S., & Li, E. (1998b). Cloning and characterization of a family of novel mammalian DNA (cytosine-5) methyltransferases. *Nature Genetics*, *19*, 219–220.

Paroush, Z., Keshet, I., Yisraeli, J., & Cedar, H. (1990). Dynamics of demethylation and activation of the α-actin gene in myoblasts. *Cell*, *63*, 1229–1237.

Pierfelice, T., Alberi, L., & Gaiano, N. (2011). Notch in the vertebrate nervous system: an old dog with new tricks. *Neuron*, *69*, 840–855.

Rai, K., Huggins, I. J., James, S. R., Karpf, A. R., Jones, D. A., & Cairn, B. R. (2008). DNA demethylation in zebrafish involves the coupling of a deaminase, a glycosylase, and Gadd45. *Cell*, *135*, 1201–1212.

Ramsahoye, B. H., Biniszkiewicz, D., Lyko, F., Clark, V., Bird, A. P., & Jaenisch, R. (2000). Non-CpG methylation is prevalent in embryonic stem cells and may be mediated by DNA methyltransferase 3a. *Proceedings of the National Academy of Sciences of the United States of America*, *97*, 5237–5242.

Razin, A., & Riggs, A. D. (1980). DNA methylation and gene function. *Science*, *210*, 604–610.

Razin, A., Szyf, M., Kafri, T., Roll, M., Giloh, H., Scarpa, S., et al. (1986). Replacement of 5-methylcytosine by cytosine: a possible mechanism for transient DNA demethylation during differentiation. *Proceedings of the National Academy of Sciences of the United States of America*, *83*, 2827–2831.

Riggs, A. D. (1975). X inactivation, differentiation, and DNA methylation. *Cytogenetics and Cell Genetics*, *14*, 9–25.

Romanov, G. A., & Vanyushin, B. F. (1981). Methylation of reiterated sequences in mammalian DNAs: effects of the tissue type, age, malignancy and hormonal induction. *Biochimica et Biophysica Acta*, *653*, 204–218.

Salvador, J. M., Brown-Clay, J. D., & Fornace, A. J., Jr. (2013). Gadd45 in stress signaling, cell cycle control, and apoptosis. In D. A. Liebermann, & B. Hoffman (Eds.), *Gadd45 stress sensor genes* (pp. 1–19). New York: Springer Science+Business Media.

Stein, R., Gruenbaum, Y., Pollack, Y., Razin, A., & Cedar, H. (1982). Clonal inheritance of the pattern of DNA methylation in mouse cells. *Proceedings of the National Academy of Sciences of the United States of America*, *79*, 61–65.

Sultan, F. A., Wang, J., Tront, J., Liebermann, D. A., & Sweatt, J. D. (2012). Genetic deletion of gadd45b, a regulator of active DNA demethylation, enhances long-term memory and synaptic plasticity. *The Journal of Neuroscience: the Official Journal of the Society for Neuroscience, 32*, 17059–17066.

Sutter, D., & Doerfler, W. (1980). Methylation of integrated adenovirus type 12 DNA sequences in transformed cells is inversely correlated with viral gene expression. *Proceedings of the National Academy of Sciences of the United States of America, 77*, 253–256.

Tahiliani, M., Koh, K. P., Shen, Y., Pastor, W. A., Bandukwala, H., Brudno, Y., et al. (2009). Conversion of 5-methylcytosine to 5-hydroxymethylcytosine in mammalian DNA by MLL partner TET1. *Science, 324*, 930–935.

Toth, M., Mueller, U., & Doerfler, W. (1990). Establishment of de novo DNA methylation patterns: transcription factor binding and deoxycytidine methylation at CpG and non-CpG sequences in an integrated adenovirus promoter. *Journal of Molecular Biology, 214*, 673–683.

Vanyushin, B. F., Alexandrushkina, N. I., & Kirnos, M. D. (1988). N^6-methyladenine in mitochondrial DNA of higher plants. *FEBS Letters, 233*, 397–399.

Vanyushin, B. F., & Belozersky, A. N. (1959). The nucleotide composition of the deoxyribonucleic acids of higher plants. *Doklady Akademii Nauk SSSR, 129*, 944–946.

Vanyushin, B. F., Belyaeva, N. N., Kokurina, N. A., Stelmashchyuk, V. Y., & Tikhonenko, A. S. (1970). Characteristics of an uracil containing DNA of phage AR 9 *Bacillus subtilis. Molecular Biology (Moscow), 4*, 724–729.

Vanyushin, B. F., & Kirnos, M. D. (1974). The nucleotide composition and pyrimidine clusters in DNA from beef heart mitochondria. *FEBS Letters, 39*, 195–199.

Vanyushin, B. F., Mazin, A. L., Vasilyev, V. K., & Belozersky, A. N. (1973). The content of 5-methylcytosine in animal DNA: the species and tissue specificity. *Biochimica et Biophysica Acta, 299*, 397–403.

Vanyushin, B. F., Nemirovsky, L. E., Klimenko, V. V., Vasiliev, V. K., & Belozersky, A. N. (1973). The 5-methylcytosine in DNA of rats: tissue and age specificity and the changes induced by hydrocortisone and other agents. *Gerontologia (Basel), 19*, 138–152.

Vanyushin, B. F., Tkacheva, S. G., & Belozersky, A. N. (1970). Rare bases in animal DNA. *Nature, 225*, 948–949.

Vanyushin, B. F., Tushmalova, N. A., & Guskova, L. V. (1974). Brain DNA methylation as an index of genome participation in mechanisms of memory formation. *Doklady Akademii Nauk SSSR, 219*, 742–744.

Vanyushin, B. F., Tushmalova, N. A., Guskova, L. V., Demidkina, N. P., & Nikandrova, L. R. (1977). The DNA methylation levels change in the rat brain on conditioned reflex elaboration. *Molecular Biology (Moscow), 11*, 181–187.

Varley, K. E., Gertz, J., Bowling, K. M., Parker, S. L., Reddy, T. E., Pauli-Behn, F., et al. (2013). Dynamic DNA methylation across diverse human cell lines and tissues. *Genome Research, 23*, 555–567.

Waalwijk, C., & Flavell, R. A. (1978). DNA methylation at a CCGG sequence in the large intron of the rabbit β-globin gene: tissue-specific variations. *Nucleic Acids Research, 5*, 4631–4642.

Weaver, I. C. G., Cervoni, N., Champagne, F. A., D'Alessio, A. C., Sharma, S., Seck, J. R., et al. (2004). Epigenetic programming by maternal behavior. *Nature Neuroscience, 7*, 847–854.

Weaver, I. C. G., Meaney, M. J., & Szyf, M. (2006). Maternal care effects on the hippocampal transcriptome and anxiety-mediated behaviors in the offspring that are reversible in adulthood. *Proceedings of the National Academy of Sciences of the United States of America, 103*, 3480–3485.

Wigler, M., Levy, D., & Perucho, M. (1981). The somatic replication of DNA methylation. *Cell, 24*, 33–40.

Wilson, V. L., Smith, R. A., Mag, S., & Cutler, R. G. (1987). Genomic 5-methyldeoxycytidine decreases with age. *The Journal of Biological Chemistry, 262*, 9948–9951.

Woodcock, D. M., Crowther, P. J., & Diver, W. P. (1987). The majority of methylated deoxycytidines in human DNA are not in the CpG dinucleotide. *Biochemical and Biophysical Research Communications, 145*, 888–894.

Wu, H., Coskun, V., Tao, J., Xie, W., Ge, W., Yoshikawa, K., et al. (2010). Dnmt3a-dependent nonpromoter DNA methylation facilitates transcription of neurogenic genes. *Science, 329*, 444–448.

Wyatt, G. R. (1950). Occurrence of 5-methylcytosine in nucleic acids. *Nature, 166*, 237–238.

Xie, W., Barr, C. L., Kim, A., Yue, F., Lee, A. Y., Eubanks, J., et al. (2012). Base-resolution analyses of sequence and parent-of-origin dependent DNA methylation in the mouse genome. *Cell, 148*, 816–831.

Yen, R.-W. C., Vertino, P. M., Nelkin, B. D., Yu, J. J., El-Deiry, W., Cumaraswamy, A., et al. (1992). Isolation and characterization of the cDNA encoding human DNA methyltransferase. *Nucleic Acids Research, 20*, 2287–2291.

Zhang, G., Huang, H., Liu, D., Cheng, Y., Liu, X., Zhang, W., et al. (2015). N^6-Methyladenine DNA modification in *Drosophila*. *Cell, 161*, 893–906.

Ziller, M. J., Müller, F., Liao, J., Zhang, Y., Gu, H., Bock, C., et al. (2011). Genomic distribution and inter-sample variation of non-CpG methylation across human cell types. *PLoS Genetics*, 7, e1002389.

CHAPTER 2

Approaches to Detecting DNA Base Modification in the Brain

X. Li[1], W. Wei[2]

[1]University of California Irvine, Irvine, CA, United States; [2]The University of Queensland, St Lucia, QLD, Australia

METHODS FOR DETECTION OF DNA MODIFICATIONS IN THE GENOME

Chromatography and mass spectrometry

The first approach developed for detecting DNA modifications was reported more than 90 years ago when Johnson & Coghill (1925) showed the presence of the pyrimidine 5-methylcytosine (5mC) in a nucleic acid by observing the optical properties of the crystalline picrate. However, the accuracy of their report has been called into question, and this technique has not been widely recognized.

In 1947, a chromatography experiment was conducted to separate amino acids by migration with organic solvents in filter paper (Consden, Gordon, & Martin, 1947). Hotchkiss (1948) then optimized this technology with a butyl alcohol system, which allows isolated nucleic acids to be recovered under favorable conditions. This, in turn, allowed purine and pyrimidine to be identified, and their quantities can be determined by UV spectrophotometry. Wyatt successfully used paper chromatography to measure the global amount of 5mC by UV spectrophotometry. Paper chromatography has thus been considered as the first method that can be used to determine DNA modifications (Wyatt & Cohen, 1952). This approach is still widely used.

After the development of improved technologies, high-performance liquid chromatography (HPLC) methods have been used to accurately assay genome-wide DNA methylation. However, HPLC methods require a large quantity (5–50 μg) of genomic DNA and synthesis of ^{32}P-labeled deoxyribonucleosides, as well as a relatively long running time (Gama-Sosa et al., 1983; Wagner & Capesius, 1981). The development of electrospray ionization (ESI) enabled liquid chromatography-mass spectrometry (LC/MS) to be used for the quantitative determination and structural characterization of polar and ionic molecules, such as nucleic acids. In 2002, a new online LC/MS method for the measurement of methylated cytosine was developed and named LC/ESI-MS. In this approach DNA is enzymatically hydrolyzed and DNA hydrolyzates subsequently are separated by reverse-phase HPLC. By coupling UV spectra analysis and the mass spectra

DNA Modifications in the Brain
ISBN 978-0-12-801596-4
http://dx.doi.org/10.1016/B978-0-12-801596-4.00002-2

of chromatographic peaks, the system can rapidly, precisely, and selectively quantify global DNA methylation. Moreover, this approach has been successfully performed on as little as 1 µg of DNA (Friso, Choi, Dolnikowski, & Selhub, 2002).

To date, merging chromatography and MS has become a common approach, allowing very sensitive and reliable quantitative detection of different DNA modifications at the whole-genome level. This approach is capable of detecting both common and rare modifications, such as 5mC, 5-hydroxymethylcytosine (5hmC), N^6-methyldeoxyadenosine (m6dA), 8-oxo-7,8-dihydroguanine, and N^7-methylguanine (Chao, Wang, Yang, Chang, & Hu, 2005; Friso et al., 2002; Greer et al., 2015; Hu, Chen, Hsu, Yen, & Chao, 2015). A major limitation of this approach is that, in general, it is not informative about the sequence context in which the DNA modifications occur. It is thus of limited utility in studying the function of any specific DNA modification.

Southern blot

Southern blot was the first method that was used in the detection of a specific DNA sequence in various DNA samples. It combines transfer of electrophoresis-separated DNA fragments to a filter membrane and subsequent fragment detection by probe hybridization. Southern blotting methods can enable the investigation of DNA modifications in a sequence context–dependent manner. For example, by using cloned X chromosome–specific probes and Southern blot, Wolf & Migeon (1982) studied methylation along 28 kb of human X chromosome DNA.

In 1978, a restriction enzyme approach was described to study 5mC at CpG dinucleotides sites that round rDNA. This work identified restriction enzymes (including HpaII, AvaI, HhaI and HaeII) that can distinguish erythrocyte rDNA from amplified DNA. The difference is attributed to the presence of 5mC rendering many restriction sites in somatic rDNA, which protects rDNA from nuclease attack. Then, by using the cloning probes that target specific genomic locus, Southern blot hybridization had an improved resolution of detection within the range of a few 100 bp (Bird, 1978).

Compared with chromatography, the Southern blot approach increases the detection resolution to a level that may be useful for examining specific genomic loci. However, it still requires a large amount of starting DNA, and the efficiency of restriction enzyme activity could affect the detection accuracy.

BISULFITE SEQUENCING FOR THE DETECTION OF 5mC

Bisulfite sequencing was first demonstrated by Frommer et al. (1992), who used it to positively identify 5mC residues in sperm DNA. The DNA was treated with sodium bisulfite and then amplified by PCR using two sets of strand-specific primers to yield amplified fragments in which thymine took the place of cytosine. Only positions at which 5mC is present and protected are amplified and the PCR amplicons then directly sequenced, with each pool representing a single molecule of the original genomic DNA. Unmodified cytosine residues that were converted to uracil will be read as thymidine during sequencing, whereas methylated cytosine is read as cytosine. Strand-specific average sequences were used to produce methylation maps, showing how frequently each cytosine residue within the amplified sequence was methylated in the original population of genomic DNA molecules.

Traditional bisulfite sequencing relied on Sanger sequencing to produce methylation maps of a single PCR amplicon. Modern next-generation sequencing techniques are exponentially faster and cheaper, enabling the sequencing of entire bisulfite-treated genomes. Using these technologies, it is feasible to conduct large-scale and high-throughput epigenetic investigations, such as comparison of average methylation states at every cytosine residue across the genome after experimental manipulation. The application of high-throughput sequencing to interrogate epigenetic modifications is commonly called epigenomics.

ANALYSIS OF BISULFITE-TREATED DNA WITHOUT SEQUENCING

Although DNA sequencing is a common technology of choice for analyzing DNA methylation after bisulfite treatment, other methods have been developed, which may be cheaper and more flexible. The most widely used technique is methylation-specific PCR. After sodium bisulfite treatment, PCR primers designed based on the reference genome sequence will no longer be complementary to treated DNA; thus, methylated DNA would alter the PCR result (Herman, Graff, My H Nen, Nelkin, & Baylin, 1996). In 2007, a methylation-sensitive variant of the high-resolution melting (MS-HRM) technique was developed (Wojdacz, Alexander, & Lise Lotte, 2008). HRM is a quantitative PCR-based technique that was initially designed to detect single-nucleotide polymorphisms (SNPs), and can be easily performed by quantitative PCR. The PCR amplicon is analyzed directly by slowly ramping the temperature while measuring the decrease in fluorescence resulting from dissociation of an intercalating fluorescent dye (SYBR Green). The denaturation temperature of DNA is sequence specific, and small sequence changes, such as SNPs, or the replacement of cytosines with thymines in bisulfite-treated DNA, can be measured through their effect on melting temperature. This technique allows direct quantitation in a single-tube assay, shorter detection duration, and cheaper price. However, MS-HPM assesses methylation profile in the amplified region as a whole rather than at specific nucleotides.

Gonzalgo & Gangning (2007) developed a rapid quantitation method called methylation-sensitive single nucleotide primer extension (MS-SNuPE). In MS-SNuPE, DNA is converted by bisulfite treatment and bisulfite-specific primers are annealed to the genomic location up to the base pair immediately before the CpG of interest. Then, DNA polymerase is introduced alongside radioactive dideoxynucleotides, permitting extension of the primer by one base only, onto the residue (C or T) present at the CpG site of interest. The C-to-T ratio can be used to quantitatively determine the percentage of genomic DNA molecules that were methylated at this specific site. This method has been improved by using ion pair reverse-phase HPLC to distinguish primer extension products to measure methylation level (Matin, Alessandra, & Hornby, 2002; Robinson et al., 1997).

To further take advantage of bisulfite conversion, Mathias et al. (2005) developed another approach by adding a base-specific cleavage step to enhance the information gained from the nucleotide changes. First, DNA is treated by bisulfite; treated DNA then undergoes in vitro transcription, generating complementary RNA fragments. Ribonuclease A is then used to cleave RNA specifically at cytosine and uracil ribonucleotides. The cleaved fragments are then analyzed by matrix-assisted laser desorption ionization/time of flight; methylation levels are measured by C-to-U conversions in forward strand or shift in fragment mass by G-to-A conversions in the amplified reverse strand. This new approach is efficient enough for high-throughput screening and interrogating numerous CpG sites.

EXTENDING BISULFITE SEQUENCING: BEYOND 5mC

Kriaucionis & Heintz (2009) described the presence of 5hmC in the mammalian genome. The discovery of this hydroxylated form of 5mC and the ten-eleven translocation (Tet) family of enzymes required for its conversion was a step toward revealing the true complexity of DNA modifications in the genome (Tahiliani et al., 2009). Subsequent research has revealed the existence of two more modified forms of cytosine. Tet enzymes can oxidize 5mC to 5hmC, then to 5-formylcytosine (5fC), 5-carboxylcytosine (5caC), and finally to unmodified cytosine (Ito et al., 2011). (This is referred to as the "classic DNA demethylation pathway.") 5mC and 5hmC are both protected from conversion to uracil during bisulfite treatment, whereas 5fC and 5caC are not protected; standard bisulfite treatment is therefore unable to differentiate between 5mC, 5hmC, 5fC, and 5caC (Huang et al., 2010). The revelation of these new modified DNA bases has necessitated refinement of the bisulfite sequencing workflow to distinguish them from each other.

In 2012, two laboratories (from the Babraham Institute and the University of Cambridge) demonstrated a method called oxidative bisulfite sequencing (oxBS-seq). Booth et al. developed this 5hmC conversion technique that selectively oxidizes 5hmC to 5fC, which can be converted to uracil during subsequent bisulfite treatment, whereas 5mC remains protected (Boch et al., 2009). By simultaneously performing bisulfite sequencing and oxBS-seq and then comparing the results, it becomes possible to distinguish between 5mC and 5hmC in genomic DNA with single-base resolution. Only 3 months later, the laboratory of Chuan He (at the University of Chicago) developed a different approach, named Tet-assisted bisulfite sequencing of 5-hydroxymethylcytosine (TAB-seq). TAB-seq uses β-glucosyltransferase to protect 5hmC from Tet1-mediated oxidation; 5mC is then converted to 5caC. After subsequent bisulfite treatment and PCR amplification, both cytosine and 5caC are changed to thymine and only 5hmC reads as C; again, the result is interpreted by comparing it to a simultaneous standard bisulfite sequencing experiment (Miao et al., 2012). Furthermore, Urich, Nery, Lister, Schmitz, & Ecker (2015) demonstrated a modified genome-wide bisulfite sequencing called MethylC-sequencing, which could be able to separate 5mC from 5hmC.

Although we now have two different approaches to differentiate 5mC from 5hmC, neither is currently useful to detect naturally occurring 5fC and 5caC. In addition, limitations inherent to the bisulfite treatment technique limit its usefulness for many applications. First, the accuracy of bisulfite treatment can be affected by incomplete conversion, where not all of the unmethylated cytosines in a sample are converted to uracil during treatment, and they are thus incorrectly interpreted as methylated cytosines during downstream analysis. The reaction parameters (such as temperature and salt concentration) are critical for bisulfite treatment to work efficiently (Fraga & Manel, 2002; Olek, Oswald, & Walter, 1996). Moreover, bisulfite treatment is technically demanding and time-intensive, with a typical experiment requiring 7–14 days of bench time (Boch et al., 2009; Miao et al., 2012). Furthermore, the harsh nature of the chemical treatment results in degradation of the starting material. Almost 90% of the total input DNA is degraded by the bisulfite conversion step, and conditions likely to ensure complete conversion of unmethylated cytosines (high bisulfite concentration, elevated temperature, and long incubation times) also increase sample degradation. The difficulty of counterbalancing these limitations can result in false-positive results for methylation and poor reproducibility (Grunau, Clark, & Rosenthal, 2001). Last, many studies using bisulfite sequencing have not taken into account the presence of other forms of modified cytosine in the genome; even when they are considered, degradation of input DNA is exacerbated by the additional steps (either chemical or enzymatic) necessary to distinguish 5hmC from 5mC (Huang et al., 2010). The very large amount of input DNA, and the long processing time (Boch et al., 2009; Frommer et al., 1992; Miao et al., 2012) have led many researchers (especially those working in single-cell systems) to move away from bisulfite sequencing and seek alternative strategies to assay covalent DNA modifications.

RESTRICTION ENZYMES FOR DNA MODIFICATIONS

Restriction enzymes are one of the easiest approaches to detect modified DNA at specific genomic sites. Cleavage of DNA by a restriction enzyme may be blocked or impaired when a particular base in the recognition site is modified. For example, MspI and HpaII recognize the same sequence (CCGG); however, they are sensitive to different modification status: when the external C in the sequence CCGG is methylated, MspI and HpaII cannot cleave. Unlike HpaII, MspI can cleave the sequence when the internal C residue is methylated (Bird & Southern, 1978). Another enzyme, Pvurts1I, only cleave the sequence $^{hm}CN_{11-12}/N_{9-10}G$, which contains 5hmC (Asgar Abbas, Monika, Honorata, & Matthias, 2014; Evelina & Giedrius, 2014; Sun et al., 2015). The combination of DpnI and DpnII is use to detect m6dA; both recognize the consensus sequence GATC, but only DpnI will cleave at this site if the adenine is methylated (Fu et al., 2015; Greer et al., 2015; Heyn & Esteller, 2015; Ratel, Ravanat, Berger, & Wion, 2006). Thus, using different restriction enzymes, we could detect DNA modification beyond familiar cytosine, including 5mC and 5hmC. In addition, this approach is cost-effective and fast; however, it is still limited by the number and distribution of restriction sites in the genome. Dai et al. (2002) found that a maximum of 4100 sites can be accessed by the restriction enzymes known to be DNA modification sensitive, and specific sites of interest that are not located at restriction sites cannot be investigated using this method.

DNA IMMUNOPRECIPITATION

DNA immunoprecipitation is a large-scale enrichment technique in molecular biology that is used to isolate DNA fragments that harbor a specific modification. Antibodies raised against a specific modified purine or pyrimidine are used to capture that base from a pool of genomic DNA molecules, and its genomic context can then be investigated.

To perform DNA immunoprecipitation, purified DNA is sonicated, causing it to shear into random fragments with an average size of 500 bp (Jacinto, Ballestar, & Esteller, 2007). Depending on the characteristics of the antibody, sheared DNA may or may not be denatured and is then incubated with antibody against the specific modification of interest. The classic immunoprecipitation technique is then used: magnetic beads are used to pull the primary antibody with bound DNA fragments. After removing unbound DNA from supernatant, proteinase K is used to release DNA from antibody and magnetic beads, and DNA is then collected. Various methods, including sequencing and quantitative PCR, can be used to search for loci rich in the modification of interest.

In 2005, this technology was first described to identify sites of differential DNA methylation in normal and transformed human cells by using an antibody against 5mC (Weber et al., 2005). Thereafter, antibodies were developed that detect other modified DNA bases; for example, antibody against 5hmC, 5fC, 5caC, 6md, and several others (Fu et al., 2015; Inoue, Shen, Dai, He, & Zhang, 2011; Li et al., 2014b; Pfaffeneder et al., 2011; Szwagierczak, Bultmann, Schmidt, Spada, & Leonhardt, 2010). Therefore, one of the main advantages of this technique is that it is not limited to 5mC, 5hmC, or both, nor is it greatly affected by the chemical structure or reactivity of any particular DNA modification. Moreover, it does not require any chemical conversation, thereby eliminating concerns about incomplete treatment and vastly reducing the required time and quantity of input DNA. In addition, since shearing of the input DNA by sonication is random, sequence biases are not a concern. The primary limitation of immunoprecipitation is the quality of antibodies; modified DNA bases share a similar chemical structure with the parent base and each other, making cross-reactivity a significant concern. Furthermore, the resolution of detection is limited by the fragment size; it is not possible to determine where within the 500-base captured fragment the target modification occurred.

An alternative form of DNA immunoprecipitation uses a protein affinity approach instead of antibody; its utility is limited to detecting 5mC with a CpG context. Modified proteins carrying a methyl-binding domain (MBD) are added to the sample in place of the antibody, where they bind to methylated DNA, which is then precipitated. Normally, MBD enrichment uses MBD 2b from MeCP2 (Brinkman et al., 2010), which has a higher affinity for double-stranded, methylated DNA and it is further enhanced by the addition of MBD3L1 (Rauch & Pfeifer, 2010). This technique avoids concerns about antibody specificity and binding sequence (Boch et al., 2009); however, it suffers from the same limitations regarding fragment size, strand specificity, and resolution.

OTHER CHEMICAL APPROACHES

Great progress has been made in neurobiology through the application of chemistry to reveal the molecular basis for behavior and the properties of the genome. Chemical treatments have been indispensable for the detection of changes on DNA, particularly for detection of rare DNA modifications (such as 5fC and 5caC) at single-base resolution.

Although bisulfite treatment remains the most widely used chemical treatment to detect modified DNA bases, other approaches have been characterized and many are able to detect modifications to bases other than 5mC and 5hmC. Lu et al. (2013) first demonstrated chemical labeling to detect 5caC. Then, Hardisty, Kawasaki, Sahakyan, and Balasubramanian (2015) successfully used a selective chemical labeling to study modified uracil bases (5-hydroxymethyluracil and 5-formyluracil) in the genome. A Chinese group discovered that treatment of DNA that with hot piperidine could produce a specific cleavage at the position of 5fC, providing a convenient method to selectively detect 5fC (Mao et al., 2013); it is similar to restriction enzyme–based methods, but has no sequence bias. Other groups have developed alternative methods to assay 5fC; Xia et al. (2015) invented a bisulfite-free method for whole-genome analysis of 5fC based on selective chemical labeling of 5fC and subsequent C-to-T transition during PCR, which could obtain genome-wide maps of 5fC at single-base resolution.

SUMMARY

Over the past several decades, recognition has grown within the scientific community that DNA encodes information in other forms than merely the nucleotide sequence. All four canonical DNA bases—adenosine, guanosine, thymidine, and cytosine—are known to occur in covalently modified forms within the genome, and nonstandard bases, such as inosine and uracil (Alseth, Dalhus, & Bjørås, 2014; Guo, Su, Zhong, Ming, & Song, 2011b; Hardisty et al., 2015), have also been identified. These modified bases can control the way the genome is packaged, accessed, and interpreted by cellular machinery, and they likely have essential roles in transcriptional regulation.

More than 90 years of research has contributed to our understanding of how and why these bases occur in the genome. In the last decade, the availability of affordable high-throughput sequencing and the advent of epigenomics workflows have vastly increased our capacity to interrogate these modifications at the genome-wide level. Emerging technologies are making it possible to study DNA modifications by using a low amount of input material, and exciting questions are now being asked about the role of modified DNA bases in very specific tissue types or even single cells.

Many enduring questions within neuroscience come back to the observation that the brain is able to respond to the environment and adapt and learn through a process of change, which is lifelong. The field of epigenomics represents an attempt to address this observation by considering the ways in which neurons might encode information without altering the underlying nucleotide sequence. In the next section, we address the application of epigenomic techniques in the context of neuroscience, to study covalent DNA modifications in the brain.

DETECTION OF DNA MODIFICATIONS IN THE BRAIN

Since 1925, 5mC was found as a minor base in various genomes (Johnson & Coghill, 1925). However, it was not identified in the brain until 1974, when Vanyushin, Nemirovsky, Klimenko, Vasiliev, & Belozersky (1972), using a thin-layer chromatography method, quantified methylated DNA in neocortex, hippocampus, and cerebellum from adult rats. Using a light conditioned reflex model, they determined a global increase in 5mC in hippocampus and neocortex; thus, this study provided the first evidence that DNA modification is involved in learning and memory processes. Surprisingly, the role of DNA methylation in modulating behavior was not investigated again for almost three decades.

CANDIDATE GENE APPROACH

Weaver et al. (2004) implicated DNA methylation was involved in modulation of adult stress responses by early life experiences. In rats, several behavioral characteristics of good mothering, including arched-back nursing (ABN) and licking and grooming (LG) of the offspring, are displayed with varying frequency by different dams. Offspring of attentive mothers exhibit higher expression of the glucocorticoid receptor (GR) in the hippocampus as adults, and they are better able to cope with stress. In contrast, low levels of maternal attention as measured by ABN and LG behavior are associated with reduced expression of the GR in hippocampus, and reduced ability to cope with stress, in adulthood. By using bisulfite sequencing, Weaver and colleagues mapped the methylation level of individual CpG site within the exon 1_7 promoter of the GR gene. When they compared results from the two cohorts of offspring, they found that rats with less attentive mothers exhibited higher level of methylation at 15 different CpG sites within the exon 1_7 promoter, resulting in reduced expression of the GR gene and abnormal behavior. This has become the first definitive evidence that site-specific CpG methylation could arise in response to experience or the environment and could modulate behavior. The same approach was later used to define a cortical DNA methylation pattern at specific genomic loci contributes to maintenance of remote memory in adult rats (Miller et al., 2010; Miller & Sweatt, 2007).

Because of massive reductions in the cost of genome-wide sequencing approaches, after 2010, few studies have moved away from the candidate gene approach in search of higher output, less biased methods to identify differentially methylated genes.

MICROARRAY SCREENING

The microarray platform has been extensively used to survey methylation in the brain. The most well-known methylation microarray is Infinium Human Methylation 450K from Illumina, which covers 99% of Refseq genes and 95% of CpG islands with additional coverage in "island shores" and the regions flanking them. Furthermore, it includes promoter regions, the 5′/3′ untranslated region, first exon, gene body, and CpG islands within the intergenic region and microRNA promoter regions (Carless, 2014). With this comprehensive coverage of human genome, several studies used this microarray for epigenome-wide association studies with brain and psychiatric diseases (Humphries et al., 2015; Kinoshita et al., 2013; Moore, McKnight, Craig, & O'Neill, 2014; Song et al., 2014). Those studies have mostly sought to reveal differences in DNA methylation between individuals that are associated with susceptibility to psychiatric illness, or to illuminate the role of DNA methylation in memory formation and maintenance. The disadvantage of using microarrays for these applications is that they only detect methylation changes in preselected regions of a specific genome. Human Methylation 450K is the only methylation microarray product currently available, and the utility of this method is thus limited to human DNA samples.

An increasing number of laboratories are choosing to avoid these issues by moving directly to whole-genome sequencing, which is increasingly affordable. Whole-genome sequencing is more flexible because it can be applied to any species and is an unbiased approach to detecting DNA modifications. More importantly, next-generation sequencing technology has increased the volume and speed of processing, and the throughput of sequencing is substantially higher than that of microarrays.

GENOME-WIDE SEQUENCING APPROACHES

Rollins (2005) directly sequenced 2565 methylated domains from human brain DNA by combining enzymatic digestion followed genome-wide sequencing. This is the first study using sequencing approach to study DNA methylation in a large-scale structure of neuronal genome. Merging with studies from other fields, the majority of neuroscientists were thinking DNA methylation in brain is a stable DNA modification, until Guo et al. (2011a, 2014) revealed methylome in postnatal brains is dynamicly regulated by neuronal activation by using genome-wide bisulfite sequencing. Then, Lister et al. reported a comprehensive profile of 5mC in human and mouse frontal cortex throughout their life span. Even more interestingly, the data from genome-wide sequencing reveals that non–CpG cytosine methylation is much more common in the brain than in any other adult tissue type (Guo et al., 2014; Lister et al., 2013). These findings have led the field of neuroscience to the realization that DNA methylation is far more dynamic and complex than initially expected and that a balance between DNA methylation and DNA demethylation is constantly fine-tuned to maintain gene expression networks in the brain.

Although indispensable for profiling whole-genome methylation at the level of the entire adult brain, there is an increasing understanding that bisulfite sequencing is not an appropriate approach to detect DNA methylation in the brain. Findings relating 5hmC, Tet proteins, and active DNA demethylation to neuronal function have made it clear that a detection method that cannot identify 5fC and 5caC will lose a great deal of information. Also, the requirement of large amount of input DNA makes it impossible to study tissues that are too small to yield the necessary DNA amount, such as small regions of the rodent brain, or even single neurons, in addition to other limitations we mentioned previously. DNA immunoprecipitation is becoming a more common approach to detect dynamic DNA modifications, such as 5mC, 5hmC, 5fC, and 5caC, under neuronal activities. Li et al. (2014b) developed a novel strategy for genome-wide sequencing based on DNA immunoprecipitation. A quantity of input DNA as small as 50 ng was individually barcoded and then mice were pooled. Through this technique, only 50 ng of input DNA can provide a reliable, comprehensive analysis of different DNA modification patterns across the entire neuronal genome. This would theoretically permit genome-wide DNA modification profiling from as low as 10,000 cells. To further extend this new sequencing approach, fluorescence-activated cell sorting has been used to separate neurons (NeuN$^+$) from glial cells and other cell types (NeuN$^-$) of the ventromedial prefrontal cortex of individual adult C57BL/6 mice (Li, Baker-Andresen, Zhao, Marshall, & Bredy, 2014a). These studies described distinct differences on methylation pattern between neurons and nonneuronal cells. It is thus important to perform experiments using isolated neurons, rather than heterogenous brain tissue, to accurately profile the role of DNA modifications in information storage within the neuronal genome. This approach has been used to successfully provide a tissue- and cell-specific DNA methylation map during cocaine self-administration (Baker-Andresen et al., 2015).

CONCLUSIONS AND FUTURE DIRECTIONS

There are at least 20 more covalent DNA modifications, occurring on all four canonical bases, and the majority are known to occur in the genome of higher eukaryotes (David, O'Shea, & Kundu, 2007). Technical limitations, such as the lack of sensitivity, capacity, and accuracy in traditional experimental approaches, have meant that some of these bases were only discovered in the last 10 years, and there are very likely others that have not yet been identified. Ongoing development of new technologies is bringing this avenue of research into reach for the first time.

PacBio has recently developed a new sequencing technology called single molecule real-time sequencing (SMRT-seq). With great depth of coverage (250×), SMRT-seq could enable us to detect up to 25 different base modifications (Flusberg et al., 2010). Currently, the required amount of input DNA to achieve this depth of coverage is a major limitation of the technique; however, rapid progress is being made in reducing the necessary sequencing coverage to obtain reproducible results, and it seems likely that this issue may eventually be solved.

True technological innovation in epigenomics research involves merging concepts and techniques from chemistry, biology, and physics. Within the past decade, the field of neuroscience has moved from first realizing that DNA methylation modulates gene expression and behavior in the brain to developing a variety of approaches to detect and quantify DNA modifications within the brain and to explore their effects on neuronal function. If this enormous rate of progress is sustained, we speculate that within the next decade, newly developed technologies will permit us to detect a variety of DNA modifications made to all four canonical bases, possibly at a single-cell level. We expect that these technologies would lead to identification of specific roles for at least some of these modifications in the function and modulation of the neuronal genome, bringing us closer to understanding how individual cells can collectively encode the traits, skills, and memories that make all of us who we are.

REFERENCES

Alseth, I., Dalhus, B., & Bjørås, M. (2014). Inosine in DNA and RNA. *Current Opinion in Genetics & Development*, 26(26C), 116–123.
Asgar Abbas, K., Monika, K., Honorata, C., & Matthias, B. (2014). Crystal structure of the 5hmC specific endonuclease PvuRts1I. *Nucleic Acids Research*, 42(9), 5929–5936.

Baker-Andresen, D., Zhao, Q., Li, X., Jupp, B., Chesworth, R., Lawrence, A. J., et al. (2015). Persistent variations in neuronal DNA methylation following cocaine self-administration and protracted abstinence in mice. *Neuroepigenetics*, *4*, 1–11. http://dx.doi.org/10.1016/j.nepig.2015.10.001.

Bird, A. P. (1978). Use of restriction enzymes to study eukaryotic DNA methylation: II. *Journal of Molecular Biology*, *118*(1), 27–47.

Bird, A. P., & Southern, E. M. (1978). Use of restriction enzymes to study eukaryotic DNA methylation: I. The methylation pattern in ribosomal DNA from *Xenopus laevis*. *Journal of Molecular Biology*, *118*(1), 27–47.

Boch, J., Scholze, H., Schornack, S., Landgraf, A., Hahn, S., Kay, S., et al. (2009). Breaking the code of DNA binding specificity of TAL-type III effectors. *Science*, *326*(5959), 1509–1512. http://dx.doi.org/10.1126/science.1178811.

Brinkman, A. B., Simmer, F., Ma, K., Kaan, A., Zhu, J., & Stunnenberg, H. G. (2010). Whole-genome DNA methylation profiling using MethylCap-seq. *Methods*, *52*(3), 232–236. http://dx.doi.org/10.1016/j.ymeth.2010.06.012.

Carless, M. A. (2014). Determination of DNA methylation levels using Illumina HumanMethylation450 BeadChips. *Methods in Molecular Biology*, *1288*, 143–192. http://dx.doi.org/10.1007/978-1-4939-2474-5_10.

Chao, M.-R., Wang, C.-J., Yang, H.-H., Chang, L. W., & Hu, C.-W. (2005). Rapid and sensitive quantification of urinary N7-methylguanine by isotope-dilution liquid chromatography/electrospray ionization tandem mass spectrometry with on-line solid-phase extraction. *Rapid Communications in Mass Spectrometry*, *19*(17), 2427–2432. http://dx.doi.org/10.1002/rcm.2082.

Consden, R., Gordon, A. H., & Martin, A. J. (1947). Gramicidin S: the sequence of the amino-acid residues. *The Biochemical Journal*, *41*(4), 596–602.

Dai, Z., Weichenhan, D., Wu, Y. Z., Hall, J. L., Rush, L. J., Smith, L. T., et al. (2002). An AscI boundary library for the studies of genetic and epigenetic alterations in CpG islands. *Genome Research*, *12*(10), 1591–1598.

David, S. S., O'Shea, V. L., & Kundu, S. (2007). Base-excision repair of oxidative DNA damage. *Nature*, *447*(7147), 941–950. http://dx.doi.org/10.1038/nature05978.

Evelina, Z., & Giedrius, S. (2014). Chemical display of pyrimidine bases flipped out by modification-dependent restriction endonucleases of MspJI and PvuRts1I families. *PLoS One*, *9*(12), e114580.

Flusberg, B. A., Webster, D. R., Lee, J. H., Travers, K. J., Olivares, E. C., Clark, T. A., et al. (2010). Direct detection of DNA methylation during single-molecule, real-time sequencing. *Nature Methods*, *7*(6), 461–465. http://dx.doi.org/10.1038/nmeth.1459.

Fraga, M. F., & Manel, E. (2002). DNA methylation: a profile of methods and applications. *BioTechniques*, *33*(3), 632–649.

Friso, S., Choi, S.-W., Dolnikowski, G. G., & Selhub, J. (2002). A method to assess genomic DNA methylation using high-performance liquid chromatography/electrospray ionization mass spectrometry. *Analytical Chemistry*, *74*(17), 4526–4531. http://dx.doi.org/10.1021/ac020050h.

Frommer, M., McDonald, L. E., Millar, D. S., Collis, C. M., Watt, F., Grigg, G. W., et al. (1992). A genomic sequencing protocol that yields a positive display of 5-methylcytosine residues in individual DNA strands. *Proceedings of the National Academy of Sciences of the United States of America*, *89*(5), 1827–1831.

Fu, Y., Luo, G.-Z., Chen, K., Deng, X., Yu, M., Han, D., et al. (2015). N^6-methyldeoxyadenosine marks active transcription start sites in chlamydomonas. *Cell*, 1–15. http://dx.doi.org/10.1016/j.cell.2015.04.010.

Gama-Sosa, M. A., Midgett, R. M., Slagel, V. A., Githens, S., Kuo, K. C., Gehrke, C. W., et al. (1983). Tissue-specific differences in DNA methylation in various mammals. *Biochimica et Biophysica Acta (BBA) – Gene Structure and Expression*, *740*(2), 212–219. http://dx.doi.org/10.1016/0167-4781(83)90079-9.

Gonzalgo, M. L., & Gangning, L. (2007). Methylation-sensitive single-nucleotide primer extension (Ms-SNuPE) for quantitative measurement of DNA methylation. *Nature Protocol*, *2*(8), 1931–1936.

Greer, E. L., Blanco, M. A., Gu, L., Sendinc, E., Liu, J., Aristizábal-Corrales, D., et al. (2015). DNA methylation on N. *Cell*, 1–12. http://dx.doi.org/10.1016/j.cell.2015.04.005.

Grunau, C., Clark, S. J., & Rosenthal, A. (2001). Bisulfite genomic sequencing: systematic investigation of critical experimental parameters. *Nucleic Acids Research*, *29*(13), E65-5.

Guo, J. U., Ma, D. K., Mo, H., Ball, M. P., Jang, M.-H., Bonaguidi, M. A., et al. (2011a). Neuronal activity modifies the DNA methylation landscape in the adult brain. *Nature Neuroscience*, *14*(10), 1345–1351. http://dx.doi.org/10.1038/nn.2900.

Guo, J. U., Su, Y., Shin, J. H., Shin, J., Li, H., Xie, B., et al. (2014). Distribution, recognition and regulation of non-CpG methylation in the adult mammalian brain. *Nature Neuroscience, 17*(2), 215–222. http://dx.doi.org/10.1038/nn.3607.

Guo, J. U., Su, Y., Zhong, C., Ming, G.-L., & Song, H. (2011b). Hydroxylation of 5-methylcytosine by TET1 promotes active DNA demethylation in the adult brain. *Cell, 145*(3), 423–434. http://dx.doi.org/10.1016/j.cell.2011.03.022.

Hardisty, R. E., Kawasaki, F., Sahakyan, A. B., & Balasubramanian, S. (2015). Selective chemical labeling of natural T modifications in DNA. *Journal of the American Chemical Society.* http://dx.doi.org/10.1021/jacs.5b03730 150609001250004.

Herman, J. G., Graff, J. R., Myöhänen, S., Nelkin, B. D., & Baylin, S. B. (1996). Methylation-specific PCR: a novel PCR assay for methylation status of CpG islands. *Proceedings of the National Academy of Science of the United States of America, 93*(18), 9821–9826.

Heyn, H., & Esteller, M. (2015). An adenine code for DNA: a second life for N6-Methyladenine. *Cell,* 1–4. http://dx.doi.org/10.1016/j.cell.2015.04.021.

Hotchkiss, R. D. (1948). The quantitative separation of purines, pyrimidines, and nucleosides by paper chromatography. *The Journal of Biological Chemistry, 175*(1), 315–332.

Hu, C.-W., Chen, J.-L., Hsu, Y.-W., Yen, C.-C., & Chao, M.-R. (2015). Trace analysis of methylated and hydroxymethylated cytosines in DNA by isotope-dilution LC\textendashMS/MS: first evidence of DNA methylation in *Caenorhabditis elegans. The Biochemical Journal, 465*(1), 39–47. http://dx.doi.org/10.1042/BJ20140844.

Huang, Y., Pastor, W. A., Shen, Y., Tahiliani, M., Liu, D. R., & Rao, A. (2010). The behaviour of 5-hydroxymethylcytosine in bisulfite sequencing. *PLoS One, 5*(1), e8888. http://dx.doi.org/10.1371/journal.pone.0008888.

Humphries, C. E., Kohli, M. A., Nathanson, L., Whitehead, P., Beecham, G., Martin, E., et al. (2015). Integrated whole transcriptome and DNA methylation analysis identifies gene networks specific to late-onset Alzheimer's disease. *Journal of Alzheimer's Disease: JAD, 44*(3), 977–987. http://dx.doi.org/10.3233/JAD-141989.

Inoue, A., Shen, L., Dai, Q., He, C., & Zhang, Y. (2011). Generation and replication-dependent dilution of 5fC and 5caC during mouse preimplantation development. *Cell Research, 21*(12), 1670–1676. http://dx.doi.org/10.1038/cr.2011.189.

Ito, S., Shen, L., Dai, Q., Wu, S. C., Collins, L. B., Swenberg, J. A., et al. (2011). Tet proteins can convert 5-methylcytosine to 5-formylcytosine and 5-carboxylcytosine. *Science, 333*(6047), 1300–1303.

Jacinto, F. V., Ballestar, E., & Esteller, M. (2007). Methyl-DNA immunoprecipitation (MeDIP): hunting down the DNA methylome. *BioTechniques, 44*(1) 35–passim.

Johnson, T. B., & Coghill, R. D. (1925). Researches on pyrimidines. C111. The discovery of 5-methyl-cytosine in tuberculinic acid, the nucleic acid of the tubercle bacillus 1. *Journal of the American Chemical Society, 47*(11), 2838–2844. http://dx.doi.org/10.1021/ja01688a030.

Kinoshita, M., Numata, S., Tajima, A., Shimodera, S., Imoto, I., & Ohmori, T. (2013). Plasma total homocysteine is associated with DNA methylation in patients with schizophrenia. *Epigenetics, 8*(6), 584–590. http://dx.doi.org/10.4161/epi.24621.

Kriaucionis, S., & Heintz, N. (2009). The nuclear DNA base 5-hydroxymethylcytosine is present in Purkinje neurons and the brain. *Science, 324*(5929), 929–930. http://dx.doi.org/10.1126/science.1169786.

Li, X., Baker-Andresen, D., Zhao, Q., Marshall, V., & Bredy, T. W. (2014a). Methyl CpG binding domain ultra-sequencing: a novel method for identifying inter-individual and cell-type-specific variation in DNA methylation. *Genes, Brain and Behavior, 13*(7), 721–731. http://dx.doi.org/10.1111/gbb.12150.

Li, X., Wei, W., Zhao, Q.-Y., Widagdo, J., Baker-Andresen, D., Flavell, C. R., et al. (2014b). Neocortical Tet3-mediated accumulation of 5-hydroxymethylcytosine promotes rapid behavioral adaptation. *Proceedings of the National Academy of Sciences of the United States of America, 111*(19), 7120–7125. http://dx.doi.org/10.1073/pnas.1318906111.

Lister, R., Mukamel, E. A., Nery, J. R., Urich, M., Puddifoot, C. A., Johnson, N. D., et al. (2013). Global epigenomic reconfiguration during mammalian brain development. *Science, 341*(6146), 1237905. http://dx.doi.org/10.1126/science.1237905.

Lu, X., Song, C.-X., Szulwach, K., Wang, Z., Weidenbacher, P., Jin, P., et al. (2013). Chemical modification-assisted bisulfite sequencing (CAB-Seq) for 5-carboxylcytosine detection in DNA. *Journal of the American Chemical Society, 135*(25), 9315–9317. http://dx.doi.org/10.1021/ja4044856.

Mao, W., Hu, J., Hong, T., Xing, X., Wang, S., Chen, X., et al. (2013). A convenient method for selective detection of 5-hydroxymethylcytosine and 5-formylcytosine sites in DNA sequences. *Organic & Biomolecular Chemistry*, *11*(21), 3568. http://dx.doi.org/10.1039/c3ob40447a.

Mathias, E., Nelson, M. R., Patrick, S., Marc, Z., Triantafillos, L., George, X., et al. (2005). Quantitative high-throughput analysis of DNA methylation patterns by base-specific cleavage and mass spectrometry. *Proceedings of the National Academy of Sciences of the United States of America*, *102*(44), 15785–15790.

Matin, M. M., Alessandra, B., & Hornby, D. P. (2002). An analytical method for the detection of methylation differences at specific chromosomal loci using primer extension and ion pair reverse phase HPLC. *Human Mutation*, *20*(4), 305–311.

Miao, Y., Hon, G. C., Szulwach, K. E., Chun-Xiao, S., Peng, J., Bing, R., et al. (2012). Tet-assisted bisulfite sequencing of 5-hydroxymethylcytosine. *Nature Protocols*, 7(12), 2159–2170.

Miller, C. A., Gavin, C. F., White, J. A., Parrish, R. R., Honasoge, A., Yancey, C. R., et al. (2010). Cortical DNA methylation maintains remote memory. *Nature Neuroscience*, *13*(6), 664–666. http://dx.doi.org/10.1038/nn.2560.

Miller, C. A., & Sweatt, J. D. (2007). Covalent modification of DNA regulates memory formation. *Neuron*.

Moore, K., McKnight, A. J., Craig, D., & O'Neill, F. (2014). Epigenome-wide association study for Parkinson's disease. *Neuromolecular Medicine*, *16*(4), 845–855. http://dx.doi.org/10.1007/s12017-014-8332-8.

Olek, A., Oswald, J., & Walter, J. (1996). A modified and improved method for bisulphite based cytosine methylation analysis. *Nucleic Acids Research*, *24*(24), 5064–5066.

Pfaffeneder, T., Hackner, B., Truss, M., Münzel, M., Müller, M., Deiml, C. A., et al. (2011). The discovery of 5-formylcytosine in embryonic stem cell DNA. *Angewandte Chemie (International Ed. in English)*, *50*(31), 7008–7012. http://dx.doi.org/10.1002/anie.201103899.

Ratel, D., Ravanat, J. L., Berger, F., & Wion, D. (2006). N6-Methyladenine: the other methylated base of DNA. *Bioessays*, *28*(3), 309–315. http://dx.doi.org/10.1002/bies.20342.

Rauch, T. A., & Pfeifer, G. P. (2010). DNA methylation profiling using the methylated-CpG island recovery assay (MIRA). *Methods*, *52*(3), 213–217. http://dx.doi.org/10.1016/j.ymeth.2010.03.004.

Robinson, C. A., Hayward-Lester, A., Hewetson, A., Oefner, P. J., Doris, P. A., & Chilton, B. S. (1997). Quantification of alternatively spliced RUSH mRNA isoforms by QRT-PCR and IP-RP-HPLC analysis: a new approach to measuring regulated splicing efficiency. *Gene*, *198*(1–2), 1–4.

Rollins, R. A. (2005). Large-scale structure of genomic methylation patterns. *Genome Research*, *16*(2), 157–163. http://dx.doi.org/10.1101/gr.4362006.

Song, Y., Miyaki, K., Suzuki, T., Sasaki, Y., Tsutsumi, A., Kawakami, N., et al. (2014). Altered DNA methylation status of human brain derived neurotrophis factor gene could be useful as biomarker of depression. *American Journal of Medical Genetics. Part B, Neuropsychiatric Genetics: the Official Publication of the International Society of Psychiatric Genetics*, *165B*(4), 357–364. http://dx.doi.org/10.1002/ajmg.b.32238.

Sun, Z., Dai, N., Borgaro, J., Quimby, A., Sun, D., Corrêa, I., et al. (2015). A sensitive approach to map genome-wide 5-hydroxymethylcytosine and 5-formylcytosine at single-base resolution. *Molecular Cell*, *57*(4), 750–761.

Szwagierczak, A., Bultmann, S., Schmidt, C. S., Spada, F., & Leonhardt, H. (2010). Sensitive enzymatic quantification of 5-hydroxymethylcytosine in genomic DNA. *Nucleic Acids Research*, *38*(19), e181. http://dx.doi.org/10.1093/nar/gkq684.

Tahiliani, M., Koh, K. P., Shen, Y., Pastor, W. A., Bandukwala, H., Brudno, Y., et al. (2009). Conversion of 5-methylcytosine to 5-hydroxymethylcytosine in mammalian DNA by MLL partner TET1. *Science*, *324*(5929), 930–935. http://dx.doi.org/10.1126/science.1170116.

Urich, M. A., Nery, J. R., Lister, R., Schmitz, R. J., & Ecker, J. R. (2015). MethylC-seq library preparation for base-resolution whole-genome bisulfite sequencing. *Nature Protocols*, *10*(3), 475–483. http://dx.doi.org/10.1038/nprot.2014.114.

Vanyushin, B. F., Nemirovsky, L. E., Klimenko, V. V., Vasiliev, V. K., & Belozersky, A. N. (1972). The 5-methylcytosine in DNA of rats. Tissue and age specificity and the changes induced by hydrocortisone and other agents. *Gerontologia*, *19*(3), 138–152.

Wagner, I., & Capesius, I. (1981). Determination of 5-methylcytosine from plant DNA by high-performance liquid chromatography. *Biochimica et Biophysica Acta*, *654*(1), 52–56.

Weaver, I. C. G., Cervoni, N., Champagne, F. A., D'Alessio, A. C., Sharma, S., Seckl, J. R., et al. (2004). Epigenetic programming by maternal behavior. *Nature Neuroscience*, 7(8), 847–854. http://dx.doi.org/10.1038/nn1276.

Weber, M., Davies, J. J., Wittig, D., Oakeley, E. J., Haase, M., Lam, W. L., et al. (2005). Chromosome-wide and promoter-specific analyses identify sites of differential DNA methylation in normal and transformed human cells. *Nature Genetics*, 37(8), 853–862. http://dx.doi.org/10.1038/ng1598.

Wojdacz, T. K., Alexander, D., & Lise Lotte, H. (2008). Methylation-sensitive high-resolution melting. *Nature Protocol*, 3(12), 1903–1908.

Wolf, S. F., & Migeon, B. R. (1982). Studies of X chromosome DNA methylation in normal human cells. *Nature*, 295(5851), 667–671. http://dx.doi.org/10.1038/295667a0.

Wyatt, G. R., & Cohen, S. S. (1952). A new pyrimidine base from bacteriophage nucleic acids. *Nature*, 170(4338), 1072–1073. http://dx.doi.org/10.1038/1701072a0.

Xia, B., Han, D., Lu, X., Sun, Z., Zhou, A., Yin, Q., et al. (2015). Bisulfite-free, base-resolution analysis of 5-formylcytosine at the genome scale. *Nature Methods*, 12(11), 1047–1050. http://dx.doi.org/10.1038/nmeth.3569.

CHAPTER 3

Active DNA Demethylation in Neurodevelopment

Y. Kang, Z. Wang, P. Jin
Emory University, Atlanta, GA, United States

INTRODUCTION

The brain is our most sophisticated organ, conveying information taken from both inside and outside the body to orchestrate a myriad of biological events for the entire organism, including behavior and emotion. The brain is highly specialized, with a staggering number of interconnections and a wide spectrum of cell types in well-organized layers. To ensure its high-order structure and performance of its dynamic functions, brain development from embryo to adulthood must be tightly orchestrated. Modification at C5 of cytosine has received significant attention as an important epigenetic mechanism involved in regulating a variety of cellular events. The complete cytosine modification machinery/cycle, from 5-methylcytosine (5mC) through 5-carboxylcytosine (5caC), seems to be tightly regulated. An abnormal cytosine modification pattern and/or unbalanced methylation versus demethylation can lead to various pathophysiological conditions (Baylin & Jones, 2011; Jakovcevski & Akbarian, 2012; Jones & Baylin, 2002; Ma et al., 2010; Urdinguio, Sanchez-Mut, & Esteller, 2009). All intermediates involved in cytosine C5 modification are expected to have their own epigenetic regulatory functions, but the contribution of each to epigenetic regulatory function should not be considered as completely separate, but rather as organized, collective entities (Fig. 3.1). In this review, we first revisit fundamental brain development and the principles of DNA methylation for translating the inscribed epigenetic codes to biological phenotypes. We then discuss how rapid technical improvements are conceptually changing our perspective of cytosine modification interplay encompassing 5-formylcytosine (5fC) and 5caC, with a special emphasis on epigenetic priming and brain development. Finally, we showcase some exciting epigenetic phenomena in our quest to decipher the epigenetic regulatory mechanisms in brain development.

FUNDAMENTAL BRAIN DEVELOPMENT

In mammals, the nervous system develops from ectoderm, the surface layer of the gastrula. Later in development, the mesoderm gives rise to the notochord, which releases the organizer proteins noggin and chordin. These proteins block the suppressive effects

DNA Modifications in the Brain
ISBN 978-0-12-801596-4
http://dx.doi.org/10.1016/B978-0-12-801596-4.00003-4

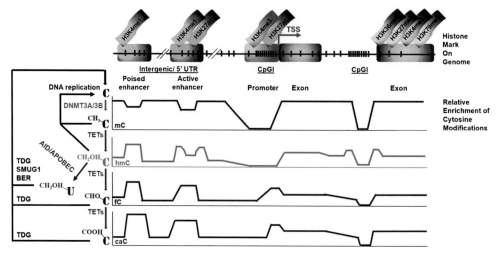

Figure 3.1 *Genomic view of distinct histone marks and cytosine modifications with schematic interplay.* Cytosines (C) are methylated by DNA methyltransferases (Dnmts) to 5-mC (CH_3-C), which are oxidized to 5-hydroxymethylcytosine (5hmC) (CH_2OH-C) by Ten-eleven translocations (TETs). TETs can oxidize the 5hmC to the 5-formylcytosine (5fC) (CHO-C) and further to 5-carboxylcytosine (5caC) (COOH-C). Both 5fC and 5caC are excised by thymine DNA glycosylase (TDG), eventually converted back to C. Genetic elements such as enhancers and promoter are defined by the histone marks and transcription factors showing characteristic C methylation. The relative enrichment pattern of each C methylation interplay element shows specific enrichments on certain genetic elements, indicating peculiar role in epigenetic regulation of gene activity. *UTR*, untranslated region.

of bone morphogenetic protein (BMP), allowing the ectoderm to form the neural plate, then the neural tube, and eventually the ventricular system, where neurogenesis proceeds within the walls of the tube to form the CNS, including the brain and spinal cord (Butler & Hodos, 2005; Siegel & Sapru, 2015). The neural plate and neural tube are composed of a single layer of neuroepithelial cells, which can be considered as neural stem cells (NSCs) (Gotz & Huttner, 2005). After closure of the neural tube, neuroepithelial cells undergo asymmetric division to generate a daughter stem cell, plus a more differentiated cell, such as a radial glial (RG) cell or a neuron (Gotz & Huttner, 2005; Huttner & Brand, 1997). With the switch to neurogenesis, all neuroepithelial cells undergo a transformation and give rise to RG cells. RG cells are fate-restricted progenitors, which can either generate nascent neurons by symmetric division or undergo self-renewal by asymmetric division (Gotz & Huttner, 2005; Yao & Jin, 2014). Both neuroepithelial cells and RG cells can generate a type of intermediate neuron progenitor cell, basal progenitors (BPs), which can generate neurons by symmetrical division (Gotz & Huttner, 2005; Haubensak, Attardo, Denk, & Huttner, 2004; Noctor, Martinez-Cerdeno, Ivic, & Kriegstein, 2004). RGs also generate astrocytes and oligodendrocytes. Some RGs remain quiescent in the subventricular zone (SVZ) and work as NSCs in adult neurogenesis (Yao & Jin, 2014).

Unlike embryonic neurogenesis, adult neurogenesis is thought to be restricted to just two regions: the SVZ of the lateral ventricle and the dentate gyrus subgranular zone (SGZ) of the hippocampus. The adult SVZ harbors radial glia–like cells (B cells), which are the SVZ stem cells. Proliferating B cells give rise to transient amplifying cells (C cells), which in turn generate neuroblasts (A cells). Through a tube formed by astrocytes, A cells form a chain, called the rostral migratory stream, and migrate toward the olfactory bulb, where the A cells are converted to different subtypes of mature neurons (Alvarez-Buylla & Lim, 2004; Ming & Song, 2011; Yao & Jin, 2014). In the SGZ, radial glia–like cells (type I cells) and nonradial precursor cells (type II cells) work as neural progenitors in the DG. These cells produce intermediate progenitors, which in turn generate neuroblasts. Neuroblasts migrate into the inner granule cell (GC) layer and differentiate into dentate GCs in the hippocampus (Ming & Song, 2011; Zhao, Deng, & Gage, 2008).

During brain development and neurogenesis (Fig. 3.2), both identity and differentiation potential are determined by orchestration between extracellular signals, such as BMP and Sonic Hedgehog (Butler & Hodos, 2005), and an intracellular network, such as transcription factors Pax6 (Balmer et al., 2012) and Dlx2 (Lim et al., 2009). Epigenetic mechanisms, including DNA methylation, histone modification, chromatin remodeling, and noncoding RNA, have been implicated in determining the DNA and histone accessibility of critical genes and in fine-tuning the expression of transcription factors. For example, Gadd45b is required for activity-induced DNA demethylation of specific promoters and the expression of corresponding genes critical for adult neurogenesis, among them brain-derived neurotrophic factor (*Bdnf*) and fibroblast growth factor-1 (*Fgf-1*) (Ma et al., 2009). Over the years, there have been interesting studies that have provided new insight into prospective epigenetic regulatory mechanisms in the nervous system. In particular, given the highly enriched level of 5-hmC in brain relative to many other tissues and cell types (for example, in Purkinje cells of the cerebellum, 5-hmC is approximately 40% as abundant as 5-mC), here we highlight the potential functional roles of this cytosine modification, and others, in brain development.

INTERPLAY OF DNA METHYLATION AND DEMETHYLATION

Initially studied in prokaryotic DNA, especially DNA cytosine MTase (Dcm), DNA adenine methylase (Dam), and cell cycle–regulated methylase (CcrM) (Palmer & Marinus, 1994; Reisenauer, Kahng, McCollum, & Shapiro, 1999), DNA methylation has become the focus of intense research in eukaryotes, especially with respect to understanding the complicated nature of cancer, stem cells, and the nervous system. In 1975, there was an early proposal that DNA methylation might be responsible for the stable maintenance of a particular gene expression pattern, but with no experimental backing (Holliday & Pugh, 1975; Riggs, 1975). Since then, enough evidence has accumulated to support this hypothesis. The identification of actively expressing genes with unmethylated promoters and

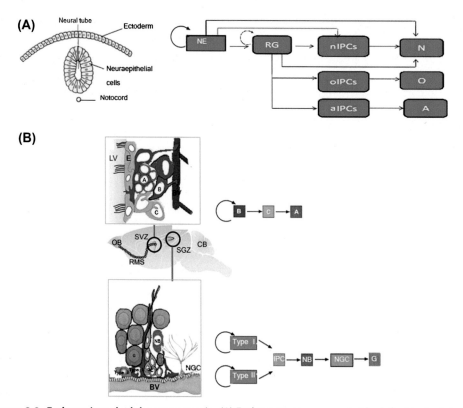

Figure 3.2 *Embryonic and adult neurogenesis.* (A) Embryonic neurogenesis. The left side depicts the structure of neural tube in early embryonic neurodevelopment stage, when the neural tube is composed of a single layer of neuroepithelial cells (NE). The right side shows the lineage of the embryonic neurogenesis. Early in CNS development, neuroepithelial cells generate more neuroepithelial cells by symmetric division. Later, they undergo asymmetric division to generate a daughter stem cell plus a more differentiated cell such as a radial glial cell (RG) or a neuron. RG can generate nascent neurons (N) either directly or via the neuronal intermediate progenitor cells (nIPC). RGs also generate astrocytes (A) and oligodendrocytes (O) via astrocyte intermediate progenitor cells (aIPC) and oligodendrocyte intermediate progenitor cells (oIPC). (B) Adult neurogenesis in subventricular zone (SVZ) and subgranular zone (SGZ). *BV*, blood vessel; *E*, ependymal cells; *G*, granule cell; *IPC*, intermediate progenitor cell; *LV*, lateral ventricle; *NB*, neuroblast; *NGC*, new granule cell; *PC*, progenitor cell; *RMS*, rostral migratory stream.

inactive genes with methylated promoters strengthened the idea (Doerfler, 1981, 1983). Furthermore, the importance of DNA methylation status for biological consequences has been established. Concomitant aberrant expression of each component of the cytosine modification machinery often causes developmental defects and pathophysiological disorders (Bestor, 2000; Chahrour et al., 2008; Guy, Cheval, Selfridge, & Bird, 2011; Lopez-Serra & Esteller, 2008; Okano, Bell, Haber, & Li, 1999). In addition, 5mC has been implicated in numerous biological phenomena, such as genetic imprinting, genetic silencing, and X chromosome inactivation (Bird, 2002; Gopalakrishnan, Van Emburgh, & Robertson, 2008; Li, 2002; Suzuki & Bird, 2008), suggesting potential key functions of DNA methylation and its associated machinery. DNA methylation is counterbalanced by DNA demethylation events, leading to a dynamic cycle of cytosine in regulating gene expression. Although passive demethylation occurs mainly through a DNA replication-dependent process, active demethylation is achieved through several players in a DNA replication–independent manner. In this section, we recapitulate key elements of cytosine modification to discuss functional roles of 5fC and 5caC since it is impossible to discuss 5fC and 5caC without exploring collectively modified cytosine interplay (Fig. 3.1).

DNA methyltransferase

In general, DNA methylation is primarily initiated and maintained through an interaction between the DNA methyltransferase (Dnmt) enzymes: DNMT1, DNMT3A, and DNMT3B. In particular, given the highly enriched level of 5-hmC in brain relative to many other tissues and cell types (for example, in Purkinje cells of the cerebellum, 5-hmC is approximately 40% as abundant as 5-mC), here we highlight the potential functional roles of this cytosine modification, and others, in brain development. It is widely accepted that DNMT3A and DNMT3B are responsible for de novo DNA methylation, whereas DNMT1 plays a major role in preserving already established patterns of DNA methylation (Chen & Li, 2006; Goll & Bestor, 2005; Li & Zhang, 2014). In preimplantation embryos, DNA double strands are largely unmethylated. In subsequent developmental stages, the expression of DNMT3A and DNMT3B is significantly augmented, and reestablishment of the DNA methylation patterns throughout the genome becomes evident; these patterns are then maintained faithfully by the function of DNMT1 together with UHRF1.

The expression of DNMTs displays distinct patterns in developing embryos and brain. Dnmt1 is found in various embryonic tissues and also expressed in most neurons, such as neural precursors and mature neurons in the adult mouse brain (Goto et al., 1994). The expression of Dnmt3b is limited to the SVZ, becoming nonexistent beyond embryonic day 15.5 (E15.5), whereas Dnmt3a is preferentially detected in the ventricular and SVZ between E10.5 and E17.5 (Feng, Chang, Li, & Fan, 2005). In the subsequent developmental stages and beyond, DNMT3a expression is abundant in most postmitotic neurons in the brain, suggesting a neuronal function for DNA methylation. Indeed, gene knockout studies have revealed its critical functional involvement in neuronal development. Due to embryonic lethality with multiple developmental abnormalities upon germline deletion of Dnmt genes (Li, Bestor, & Jaenisch, 1992; Okano et al., 1999), conditional knockout approaches have been instrumental for investigating the function of DNMTs in postnatal neurons. For example, deletion of Dnmt1 in Nestin-Cre–expressing neural stem cells leads to DNA hypomethylation and the accelerated differentiation of astrocytes (Fan et al., 2005). Emx1-Cre:Dnmt1 knockout mice, in which Dnmt1 is specifically ablated in the dorsal forebrain, display DNA hypomethylation with severe degeneration in cortex and hippocampus as well as impairment of thalamocortical long-term potentiation (Golshani, Hutnick, Schweizer, & Fan, 2005; Hutnick et al., 2009). Although CamK-Cre–mediated deletion of Dnmt1 in postmitotic neurons is compatible with normal development, double-conditional inactivation of Dnmt1 and Dnmt3a results in the generation of smaller hippocampi and impaired learning and memory, which is accompanied by reduced levels of 5mC and 5hmC (Feng et al., 2010). Dnmt3a deletion in Nestin-Cre NSCs suggests a role in neuromuscular control, as evidenced by a significant decrease in motor neurons and an abnormal structure in neuromuscular junctions of the diaphragm muscle (Nguyen, Meletis, Fu, Jhaveri, & Jaenisch, 2007). Together, these studies indicate the potentially critical roles of DNA methylation, regulated mainly by the DNMT proteins, in neuronal development and function throughout embryogenesis and adult life.

Ten-eleven translocation cytosine dioxygenase

Since the initial identification of Ten-eleven translocation (TET)1 in acute myeloid leukemia as a fusion partner of MLL, much attention has been paid to the TET family of proteins, which consists of TET1, TET2, and TET3. Following the initial report by Rao in 2009, which demonstrated TET1 can convert 5mC to 5hmC (Tahiliani et al., 2009), several studies went on to show that all three TET proteins have the ability to generate the higher oxidized forms of cytosine, including 5fC and 5caC, as well as 5hmC, in a stepwise enzymatic reaction. With recent advances in technology, we can now detect the presence of 5hmC, as well as 5fC and 5caC, at single-base resolution, and it is clear that TET family proteins represent one of the major players in regulating gene expression epigenetically via active DNA demethylation. Small interfering RNA (siRNA)–mediated knockdown of Tet1 or Tet1/Tet2 in ESCs leads to decreased expression of pluripotency genes; impaired self-renewal capability; and skewed developmental potential, along with a reduced level of 5mhC (Ficz et al., 2011; Freudenberg et al., 2012; Ito et al., 2010; Koh et al., 2011). Unexpectedly, ESCs deficient in Tet1 show no defects in pluripotency, but they have altered differentiation potential (Dawlaty et al., 2011). Mice deficient in Tet1 are born alive with some defects, including delayed body development and compromised fertility (Dawlaty et al., 2011). TET2 knockout mice display hematopoietic defects (Ko et al., 2011); and TET3, which is highly expressed in oocyte, has been implicated in epigenetic reprogramming of the paternal genome after fertilization (Gu et al., 2011). Tet1:Tet2 double knockout ESCs show unaltered pluripotency and are still compatible with embryonic and perinatal development, albeit with varying degrees of developmental abnormalities (Dawlaty et al., 2013). Again, the double knockout ESCs and mice retain reduced levels of 5hmC. Similarly, depletion of all three Tets in ESCs leads to impaired differentiation ability, with hypermethylation on promoters of genes related to early development (Dawlaty et al., 2014). In an independent study, it was shown that Tet1/2/3-deficient ESCs are capable of maintaining self-renewal and pluripotency, although Tet triple knockout mouse embryonic fibroblast cells fail to become pluripotent stem cells upon overexpression of the Yamanaka factors (Hu et al., 2014). Together, these results suggest, albeit not without controversy over TET1 function, that oxidative methylation and demethylation of cytosine is required for early development and cellular functions.

All three TET proteins are expressed in the brain (Hahn et al., 2013; Kaas et al., 2013). Upregulated expression of Tet1 is found consistently in the cortex of psychosis patients. In a subsequent detailed analysis of TET1 knockout mice, Zhang et al. (2013) revealed that a deficiency in Tet1 leads to impaired proliferation of neural progenitor cells (NPCs) in the adult brain, as well as defective learning and memory. Promoters of neurogenesis-related genes are hypermethylated and their expression decreased in Tet1−/− NPCs. Rudenko et al. (2013) reported defects in long-term potentiation and memory extinction, with an increase of methylation on the *Npas4* gene in Tet1-deficient mice. Conversely, a decrease in DNA methylation with a concomitant enhanced expression of several neuronal genes were seen upon overexpression of TET1 in the brain (Guo, Su, Zhong, Ming, & Song, 2011; Kaas et al., 2013). The role of TET2 and TET3 in NPC differentiation was uncovered by a knockdown study in the brain cortex (Hahn et al., 2013). Combined with a recent study revealing the function of TET3 in behavioral adaptation and in regulating excitatory glutamatergic synaptic transmission and plasticity (Li et al., 2015; Li et al., 2014; Yu et al., 2015), these findings clearly support the notion that TET-mediated DNA demethylation is pivotal for neuron development and postmitotic neuronal function.

Thymine DNA glycosylase

Thymine DNA glycosylase (TDG) is a member of the uracil DNA glycosylase (UDG) superfamily and has the ability to catalyze the glycosidic bond between the base and deoxyribose sugar of DNA-generating apurinic/apyrimidinic (AP) sites, which can be subjected to the base excision repair pathway. A series of enzymatic reactions, including AP endonuclease, DNA polymerase, and DNA ligase, can remove T from G·T or U bases from G·U mismatches and replace it with C in double-stranded DNA. With the emerging role of cytosine methylation and demethylation in epigenetics, recent studies have revealed that TDG can also excise 5fC from G·5fC pairs and 5caC from G·5caC pairs with high efficiency (He et al., 2011; Maiti & Drohat, 2011). Consistently, a strong reciprocal relation between the level of 5fC, 5caC, and TDG has been reported; although the amount of both 5fC and 5caC is significantly reduced upon overexpression of TDG together with TET2 (Nabel et al., 2012), there is greater augmentation of 5fC and 5caC in Tdg knockout mice (He et al., 2011; Shen et al., 2013; Song et al., 2013). Furthermore, among members of the UDG family, including UNG, MBD4, and SMUG, TDG is the only protein to exhibit an essential function for embryonic development, as demonstrated by the embryonic lethality of *Tdg*−/− mice (Cortazar et al., 2011; Cortellino et al., 2011).

Collectively, these studies point to TDG as an important protein for mediating the demethylation of 5mC.

There are several proposals to explain how 5mC can be processed in DNA demethylation pathways (Fig. 3.1). Among them, active DNA demethylation through the TET-BER (TDG) pathway is the one strongly supported by experimental results from multiple groups. In this scenario, TET proteins first generate 5hmC by oxidizing 5mC and subsequently convert 5hmC to 5fC, and then to 5caC. The resulting oxidized cytosine derivatives can be recognized and excised by TDG, leading to replacement with unmodified cytosine. Another interesting report is that such active DNA demethylation in mouse brain involves the cytosine deamination step mediated by activation-induced deaminase/apolipoprotein B mRNA-editing enzyme complex (AID/APOBEC) (Guo et al., 2011). Although this has been supported by several lines of evidence (Bhutani et al., 2010; Kumar et al., 2013; Morgan, Dean, Coker, Reik, & Petersen-Mahrt, 2004; Popp et al., 2010), other reports showed that AID/APOBEC primarily deaminates unmodified cytosine, but has significantly reduced activity on 5mC and no detectable deamination activity on 5hmC (Nabel et al., 2012). Also, the AID has a preference for single-stranded DNA in mediating cytosine deamination (Bransteitter, Pham, Scharff, & Goodman, 2003).

MAPPING AND QUANTIFICATION OF THE DERIVATIVES OF 5mC

To understand the mechanism and significance of the dynamic balance between cytosine methylation and demethylation in numerous biological processes in both normal and disease states, it is imperative to grasp collective ideas of the whole-genome–wide cytosine modification landscape and the quantification of differential enrichment or depletion of each derivative at single-base resolution. Over the past few years, researchers have endeavored to develop several techniques for mapping and quantifying the derivatives of 5mC (5hmC, 5fC, and 5caC); they have achieved striking technical advances in next-generation sequencing based on deep sequencing and grandly raised the scale of study from a single locus to the whole genome, with resolution at a single-base level. Here we discuss widely used methods in two different, but interdependent, categories: pull-down–based methods and bisulfite sequencing (BS-seq)–based methods (Fig. 3.3). Simply, the principle behind all these mapping methods is based on selective modification, enrichment, or both using specific affinity alterations, differential alterations, or a combination with chemicals or enzymes on specific residue(s) for selective readout. In general, affinity-based profiling is relatively cost-effective, but it has lower resolution and lacks collective information on the relative enrichment at each modification locus. The low resolution of affinity or pull-down–based methods and recognition of the importance of derivatives of 5mC in biological processes drive the field to develop additional technologies to map 5mC and its oxidative forms at single-base resolution, and to

	BS	oxBS	TAB	fCAB	redBS	caCAB	MAB
		KRuO$_4$	βGT, UDG-Glc, TET	EtONH$_2$	NaBH$_4$	EDC R-NH$_2$	M.SssI
C	T	T	T	T	T	T	C
5mC	C	C	T	C	C	C	C
5hmC	C	T	C	C	C	C	C
5fC	T	T	T	C	C	T	T
5caC	T	T	T	T	T	C	T
	Direct reading	Subtract from BS	Direct reading	Subtract from BS	Subtract from BS	Subtract from BS	Subtract from BS
	5mc & 5hmC	Reading out 5hmC	Reading out 5hmC	Reading out 5fC	Reading out 5fC	Reading out 5caC	Reading out 5fC/5caC

Figure 3.3 *Mapping of 5-methylcytosine, 5-hydroxymethylcytosine, 5-formylcytosine, and 5-carboxylcytosine with bisulfite sequencing and modified bisulfite sequencing methods at single-base resolution.* In the standard bisulfite (BS) sequencing, both 5-methylcytosine (5mC) and 5-hydroxymethylcytosine (5hmC) can be identified as C. In the oxidative BS (oxBS) method, KRuO$_4$ leads to oxidization of 5hmC to 5-formylcytosine (5fC), which can be read as T in the subsequent BS treatment. Thus, 5hmC from the original genomic DNAs can be determined by subtraction from the BS (ie, T from oxBS-C from the standard BS sequencing). In the TET-assisted BS (TAB) sequencing, first, β-glucosyltransferase (β-GT) can convert 5hmC to 5-glucosylmethylcytosine (5gmC). In the subsequent treatment with TET, 5mC and 5fC become 5-carboxylcytosine (5caC). Thus, upon BS treatment, the resulting C, 5caC, can be read as T, whereas only 5gmC (ie, 5hmC in the original genomic DNA) can be read as C. The signal of 5hmC can be identified by direct reading the results. In the chemical modification–assisted 5fC-assisted bisulfite (fCAB) sequencing, *O*-ethylhydroxylamine (EtONH$_2$)–treated 5fC can be protected from the subsequent BS treatment, and thus read as C. Similarly, in the reduced BS method, treatment of NaBH$_4$ can lead to reduction of 5fC to 5hmC. Thus, in the fCAB or the reduced bisulfite (redBS) sequencing, 5fC in the original genomic DNA can be identified as C. In the chemical modification–assisted bisulfate (caCAB) sequencing, 5caC treated with 1-ethyl-3-[3-dimethylaminopropyl]-carbodiimide hydrochloride (EDC) can be protected from the subsequent BS treatment, and thus read as C. In the methylation-assisted bisulfite (MAB) sequencing, unmodified C can be methylated (5mC) by the *S*-adenosyl-methionine–dependent CpG methyltransferase M.SssI. Upon BS treatment, 5fC and 5caC, not other forms, can be read as T. Signals specific for 5fC or 5caC can be determined after subtracting the standard BS results.

distinguish individual forms quantitatively. The principle of BS-seq–based methods is conversion of cytosine with methyl moiety to cytosine with selective treatment for conversion protection and subtraction of readout from BS-seq signals. Combined with the standard BS-seq, oxidative bisulfite sequencing (oxBS-seq), and Tet-assisted bisulfite sequencing (TAB-seq) provide detailed insights into the genome-wide distribution of 5hmC with single-base resolution.

Here we list only some representative reagents as examples to better clarify the procedure. Sodium bisulfite ($NaHSO_3$) is for BS-seq to specifically deaminate unmethylated cytosine, but not others. Antibodies against 5mC, 5hmC, 5fC, and 5caC are used for DNA immunoprecipitation sequencing (Ficz et al., 2011; Jin, Wu, Li, & Pfeifer, 2011; Shen et al., 2013; Stroud, Feng, Morey Kinney, Pradhan, & Jacobsen, 2011; Williams et al., 2011; Wu et al., 2011; Xu et al., 2011). For selective chemical labeling or glucosylation, periodate oxidation, biotinylation [glucosylation, periodate oxidation, biotinylation (GLIB)], or glycosylated 5-hydroxymethylcytosine (g5hmC)-binding protein 1 (JBP1), there is T4 bacteriophage β-glucosyltransferase (β-GT) for the addition of azide-modified or -unmodified glucose to 5hmC (g5hmC), JBP1 to pull down g5hmC, and biotin probe for the addition of biotin to 5hmC (Pastor et al., 2011; Raiber et al., 2012; Robertson et al., 2011; Song et al., 2013, 2011; Terragni, Bitinaite, Zheng, & Pradhan, 2012). Another great value of BS-seq is that it can provide subtractive readout in combination with a multitude of sequencing methods modified by diverse selective chemical treatments. Some of the reported methods to map 5fC and 5caC also adopted the power of the merge with BS-seq (Fig. 3.3). In the reduced bisulfite sequencing (redBS-seq) method, with the selective reduction of 5fC to 5hmC by sodium borohydride ($NaBH_4$), followed by bisulfite treatment, 5fC is read as C in the redBS-seq and T in the BS-seq. 5fC can be elucidated quantitatively at single-base resolution by redBS readout subtraction (Booth, Marsico, Bachman, Beraldi, & Balasubramanian, 2014). Booth also developed oxBS-seq, which selectively oxidizes 5hmC to 5fC with the same principle of readout subtraction (Booth et al., 2014). In TAB-seq, 5hmC is glucosylated by β-GT into g5hmC, treated with TET, converting all other derivatives to C or caC, and then g5hmC is the only C readout. Identification of 5caC throughout the genome with single-base resolution is made possible with the chemical modification-assisted bisulfite sequencing method (Lu et al., 2013). 5caC is protected from the bisulfite treatment with 1-ethyl-3-[3-dimethylamniopropyl]-carbodiimide hydrochloride and reads as C. Since 5caC reads as T in the conventional BS-seq, 5caC can be identified by subtracting C of the CAB-seq output from T of the conventional BS-seq output.

DISCUSSION

When the cells are ready, it is critical to respond promptly to both endogenous and exogenous cues by keeping the priming states of diverse biological processes in check. It is not efficient to retain surplus protein or RNA for acute responses to environmental cues. Therefore, the priming of certain cell activities epigenetically is considered a powerful mechanism for preparing cells for changes of status, such as developmental competence. Indeed, the type, strength, and duration of epigenetic priming can be delicately coordinated for the needs of the cell with divergent epigenetic gears. The identification of epigenetic priming highlights that the epigenetic state is useful for predicting regulators relevant to later stages of differentiation.

Even with these rapid advances in epigenetics, studies in the nervous system have been hampered by the limited number of cells that results from their postmitotic nature and the enormous heterogeneity of the cells in terms of types, as well as niches in the extensive neural network. However, the disadvantage of the neuroepigenetic system can be simultaneously interpreted as an advantage in that there is no passive dilution of epigenetic information, enabling clear assessment for causation between epigenetic profiles and cellular phenotypes. In addition, some epigenetic features, such as 5hmC and non–CpG methylation, stand out significantly in brain. The extreme diversity of DNA methylation patterns may indicate specific regulatory mechanisms and functions in brain compared to the relatively well-conserved histone modification system. DNA methylation could be a cellular "dimmer switch," equipped with the more stable and conserved histone modification at the ready to control the infrastructure of transcription and translation. We are still in the early stages when it comes to understanding the mechanism of variations in DNA methylation. With several outstanding questions still to be answered and insufficient information about 5fC and 5caC, for now we must extrapolate the epigenetic implications from the study of 5hmC and apply the knowledge to investigations of 5fC and 5caC.

Aside from the most efficient way of getting rid of 5hmC and salvaging C, the trace molecules 5fC and 5caC are adopted in the DNA methylation interplay, which means those trace elements have a high possibility of functionality, rather than acting as mere intermediates. Indeed, the stable nature of 5fC has recently been reported using nano-high-performance liquid chromatography-mass spectrometry/resolution mass spectrometry (Bachman et al., 2015). Within active transcription start sites, Tdg, 5fC, and 5caC were enriched, along with 5mC and 5hmC decrements (Neri et al., 2015). Based on the spatiotemporal distribution of 5fC and 5caC detectable only during certain time windows of development (Wheldon et al., 2014) and limited enrichment on specific promoters of the genes related to differentiation, it is tempting to speculate on their specific function as determinants in lineage specification in embryonic brain development. In addition, 5caC expression intensity varies from cell to cell, in contrast to the fairly even expression level of 5hmC, indicating an even more dynamic nature. 5fC showed a global genomic distribution similar to 5hmC, but varied at some protein–DNA interaction sites, implying potentially different epigenetic function(s) for 5fC versus other signatures (Sun et al., 2015). Interestingly, both 5hmC and 5fC showed almost identical distributions on histone modification sites associated with the activity of promoters and enhancers, which provides support for a close correlation between cytosine modification and histone modification in regulating transcription. In addition, 5hmCH and 5fCH seem to be more dynamic, based on the lower overlapping ratio of 5hmCH and 5fCH between two replicates versus 5hmCpG and 5fCpG. Due to the highly reactive nature of formyl moiety, 5fC is distinct from its relative cytosine modification derivatives in its impact on the solid geometry of the DNA double helix (Raiber et al., 2015). The effect of 5fC on DNA geometry can cause a shift in the propensity to base pairing and conformation, leading to a change in accessibility, recognition, and affinity of interacting proteins, such as transcription factors and repressors. A study showing that the RNA polymerase II (Pol II) interacts with 5CaC through the 5-carboxyl group, resulting in RNA Pol II pausing (Wang et al., 2015), supports its functional significance in gene regulation. However, despite the emerging evidence of the dynamic nature of 5fC and 5caC, they require further investigation.

Many outstanding questions remain about all the roles the interplay of DNA methylation, including 5fC and 5caC, might have. How extensively are genes regulated epigenetically, especially by methylation interplay? Not all promoters or enhancers are modified with 5mC, 5hmC, 5fC, or 5caC, regardless of their expression level. Given that, what other mechanisms regulate the expression of the gene? Why are the levels of 5fC and 5caC so low compared to the level of 5hmC, the substrate of Tet? Does Tdg contribute significantly to keeping the level of 5fC and 5caC low? We should still be wide open to other possibilities of other salvage pathways in the DNA methylation interplay system. Intriguingly, 5fC seems to be more dynamic in a non–CpG context. In addition, the epigenetic function of demethylation in gene bodies and enhancers awaits further investigation. Interestingly, the correlation of gene body methylation and demethylation with gene expression seems cell type specific, at least in neurons or context dependent. An unexpected anticorrelation was observed between low-density CpG methylation and gene expression throughout the whole gene body (Guo et al., 2011). Finally, the essential question is whether DNA methylation and demethylation is necessary or sufficient for the regulation of gene expression. With the tools of genetic manipulation systems, such as conditional knockout mice, Cas9, and siRNA, global and local effects of cytosine methylation and demethylation changes can be clarified. With even more advanced methods, we should be able to completely correlate the binding of potential interacting proteins with the base-resolution maps of 5mC, 5hmC, 5fC, and 5caC to reveal their epigenetic regulatory roles not only in the brain but also in broad ranges of other biological phenomena and pathophysiological conditions.

REFERENCES

Alvarez-Buylla, A., & Lim, D. A. (2004). For the long run: maintaining germinal niches in the adult brain. *Neuron, 41*(5), 683–686. pii:S0896627304001114.

Bachman, M., Uribe-Lewis, S., Yang, X., Burgess, H. E., Iurlaro, M., Reik, W., et al. (2015). 5-Formylcytosine can be a stable DNA modification in mammals. *Nature Chemical Biology, 11*(8), 555–557. http://dx.doi.org/10.1038/nchembio.1848.

Balmer, N. V., Weng, M. K., Zimmer, B., Ivanova, V. N., Chambers, S. M., Nikolaeva, E., et al. (2012). Epigenetic changes and disturbed neural development in a human embryonic stem cell-based model relating to the fetal valproate syndrome. *Human Molecular Genetics, 21*(18), 4104–4114. http://dx.doi.org/10.1093/hmg/dds239 pii:dds239.

Baylin, S. B., & Jones, P. A. (2011). A decade of exploring the cancer epigenome – biological and translational implications. *Nature Reviews Cancer, 11*(10), 726–734. http://dx.doi.org/10.1038/nrc3130.

Bestor, T. H. (2000). The DNA methyltransferases of mammals. *Human Molecular Genetics, 9*(16), 2395–2402.

Bhutani, N., Brady, J. J., Damian, M., Sacco, A., Corbel, S. Y., & Blau, H. M. (2010). Reprogramming towards pluripotency requires AID-dependent DNA demethylation. *Nature, 463*(7284), 1042–1047. http://dx.doi.org/10.1038/nature08752.

Bird, A. (2002). DNA methylation patterns and epigenetic memory. *Genes & Development, 16*, 6–21.

Booth, M. J., Marsico, G., Bachman, M., Beraldi, D., & Balasubramanian, S. (2014). Quantitative sequencing of 5-formylcytosine in DNA at single-base resolution. *Nature Chemistry, 6*(5), 435–440. http://dx.doi.org/10.1038/nchem.1893.

Bransteitter, R., Pham, P., Scharff, M. D., & Goodman, M. F. (2003). Activation-induced cytidine deaminase deaminates deoxycytidine on single-stranded DNA but requires the action of RNase. *Proceedings of the National Academy of Sciences of the United States of America, 100*(7), 4102–4107. http://dx.doi.org/10.1073/pnas.0730835100.

Butler, A. B., & Hodos, W. (2005). *Comparative vertebrate neuroanatomy, second edition* (2nd ed.). Hoboken: John Wiley & Sons, Inc.

Chahrour, M., Jung, S. Y., Shaw, C., Zhou, X., Wong, S. T., Qin, J., et al. (2008). MeCP2, a key contributor to neurological disease, activates and represses transcription. *Science, 320*(5880), 1224–1229. http://dx.doi.org/10.1126/science.1153252.

Chen, T., & Li, E. (2006). Establishment and maintenance of DNA methylation patterns in mammals. *Current Topics in Microbiology and Immunology, 301*, 179–201.

Cortazar, D., Kunz, C., Selfridge, J., Lettieri, T., Saito, Y., MacDougall, E., et al. (2011). Embryonic lethal phenotype reveals a function of TDG in maintaining epigenetic stability. *Nature, 470*(7334), 419–423. http://dx.doi.org/10.1038/nature09672.

Cortellino, S., Xu, J., Sannai, M., Moore, R., Caretti, E., Cigliano, A., et al. (2011). Thymine DNA glycosylase is essential for active DNA demethylation by linked deamination-base excision repair. *Cell, 146*(1), 67–79. http://dx.doi.org/10.1016/j.cell.2011.06.020.

Dawlaty, M. M., Breiling, A., Le, T., Barrasa, M. I., Raddatz, G., Gao, Q., et al. (2014). Loss of Tet enzymes compromises proper differentiation of embryonic stem cells. *Developmental Cell, 29*(1), 102–111. http://dx.doi.org/10.1016/j.devcel.2014.03.003.

Dawlaty, M. M., Breiling, A., Le, T., Raddatz, G., Barrasa, M. I., Cheng, A. W., et al. (2013). Combined deficiency of Tet1 and Tet2 causes epigenetic abnormalities but is compatible with postnatal development. *Developmental Cell, 24*(3), 310–323. http://dx.doi.org/10.1016/j.devcel.2012.12.015.

Dawlaty, M. M., Ganz, K., Powell, B. E., Hu, Y. C., Markoulaki, S., Cheng, A. W., et al. (2011). Tet1 is dispensable for maintaining pluripotency and its loss is compatible with embryonic and postnatal development. *Cell Stem Cell, 9*(2), 166–175. http://dx.doi.org/10.1016/j.stem.2011.07.010.

Doerfler, W. (1981). Dna methylation—a regulatory signal in eukaryotic gene expression. *Journal of General Virology, 57*(1), 1–20. http://dx.doi.org/10.1099/0022-1317-57-1-1.

Doerfler, W. (1983). Dna methylation and cene activity. *Annual Review of Biochemistry, 52*(1), 93–124. http://dx.doi.org/10.1146/annurev.bi.52.070183.000521.

Fan, G., Martinowich, K., Chin, M. H., He, F., Fouse, S. D., Hutnick, L., et al. (2005). DNA methylation controls the timing of astrogliogenesis through regulation of JAK-STAT signaling. *Development (Cambridge, England), 132*(15), 3345–3356. http://dx.doi.org/10.1242/dev.01912.

Feng, J., Chang, H., Li, E., & Fan, G. (2005). Dynamic expression of de novo DNA methyltransferases Dnmt3a and Dnmt3b in the central nervous system. *Journal of Neuroscience Research, 79*(6), 734–746. http://dx.doi.org/10.1002/jnr.20404.

Feng, J., Zhou, Y., Campbell, S. L., Le, T., Li, E., Sweatt, J. D., et al. (2010). Dnmt1 and Dnmt3a maintain DNA methylation and regulate synaptic function in adult forebrain neurons. *Nature Neuroscience, 13*(4), 423–430. http://dx.doi.org/10.1038/nn.2514.

Ficz, G., Branco, M. R., Seisenberger, S., Santos, F., Krueger, F., Hore, T. A., et al. (2011). Dynamic regulation of 5-hydroxymethylcytosine in mouse ES cells and during differentiation. *Nature, 473*(7347), 398–402. http://dx.doi.org/10.1038/nature10008.

Freudenberg, J. M., Ghosh, S., Lackford, B. L., Yellaboina, S., Zheng, X., Li, R., et al. (2012). Acute depletion of Tet1-dependent 5-hydroxymethylcytosine levels impairs LIF/Stat3 signaling and results in loss of embryonic stem cell identity. *Nucleic Acids Research, 40*(8), 3364–3377. http://dx.doi.org/10.1093/nar/gkr1253.

Goll, M. G., & Bestor, T. H. (2005). Eukaryotic cytosine methyltransferases. *Annual Review of Biochemistry, 74*, 481–514. http://dx.doi.org/10.1146/annurev.biochem.74.010904.153721.

Golshani, P., Hutnick, L., Schweizer, F., & Fan, G. (2005). Conditional Dnmt1 deletion in dorsal forebrain disrupts development of somatosensory barrel cortex and thalamocortical long-term potentiation. *Thalamus & Related Systems, 3*(3), 227–233. http://dx.doi.org/10.1017/s1472928807000222.

Gopalakrishnan, S., Van Emburgh, B. O., & Robertson, K. D. (2008). DNA methylation in development and human disease. *Mutation Research, 647*(1–2), 30–38. http://dx.doi.org/10.1016/j.mrfmmm.2008.08.006.

Goto, K., Numata, M., Komura, J. I., Ono, T., Bestor, T. H., & Kondo, H. (1994). Expression of DNA methyltransferase gene in mature and immature neurons as well as proliferating cells in mice. *Differentiation; Research in Biological Diversity, 56*(1–2), 39–44.

Gotz, M., & Huttner, W. B. (2005). The cell biology of neurogenesis. *Nature Reviews. Molecular Cell Biology, 6*(10), 777–788. http://dx.doi.org/10.1038/nrm1739 pii:nrm1739.

Gu, T. P., Guo, F., Yang, H., Wu, H. P., Xu, G. F., Liu, W., et al. (2011). The role of Tet3 DNA dioxygenase in epigenetic reprogramming by oocytes. *Nature, 477*(7366), 606–610. http://dx.doi.org/10.1038/nature10443.

Guo, J. U., Su, Y., Zhong, C., Ming, G. L., & Song, H. (2011). Hydroxylation of 5-methylcytosine by TET1 promotes active DNA demethylation in the adult brain. *Cell, 145*(3), 423–434. http://dx.doi.org/10.1016/j.cell.2011.03.022.

Guy, J., Cheval, H., Selfridge, J., & Bird, A. (2011). The role of MeCP2 in the brain. *Annual Review of Cell and Developmental Biology, 27*, 631–652. http://dx.doi.org/10.1146/annurev-cellbio-092910-154121.

Hahn, M. A., Qiu, R., Wu, X., Li, A. X., Zhang, H., Wang, J., et al. (2013). Dynamics of 5-hydroxymethylcytosine and chromatin marks in mammalian neurogenesis. *Cell Reports, 3*(2), 291–300. http://dx.doi.org/10.1016/j.celrep.2013.01.011.

Haubensak, W., Attardo, A., Denk, W., & Huttner, W. B. (2004). Neurons arise in the basal neuroepithelium of the early mammalian telencephalon: a major site of neurogenesis. *Proceedings of the National Academy of Sciences of the United States of America, 101*(9), 3196–3201. http://dx.doi.org/10.1073/pnas.0308600100 pii:0308600100.

He, Y. F., Li, B. Z., Li, Z., Liu, P., Wang, Y., Tang, Q., et al. (2011). Tet-mediated formation of 5-carboxylcytosine and its excision by TDG in mammalian DNA. *Science, 333*(6047), 1303–1307. http://dx.doi.org/10.1126/science.1210944.

Holliday, R., & Pugh, J. E. (1975). DNA modification mechanisms and gene activity during development. *Science, 187*(4173), 226–232.

Hu, X., Zhang, L., Mao, S. Q., Li, Z., Chen, J., Zhang, R. R., et al. (2014). Tet and TDG mediate DNA demethylation essential for mesenchymal-to-epithelial transition in somatic cell reprogramming. *Cell Stem Cell, 14*(4), 512–522. http://dx.doi.org/10.1016/j.stem.2014.01.001.

Hutnick, L. K., Golshani, P., Namihira, M., Xue, Z., Matynia, A., Yang, X. W., et al. (2009). DNA hypomethylation restricted to the murine forebrain induces cortical degeneration and impairs postnatal neuronal maturation. *Human Molecular Genetics, 18*(15), 2875–2888. http://dx.doi.org/10.1093/hmg/ddp222.

Huttner, W. B., & Brand, M. (1997). Asymmetric division and polarity of neuroepithelial cells. *Current Opinion in Neurobiology, 7*(1), 29–39 pii:S0959-4388(97)80117-1.

Ito, S., D'Alessio, A. C., Taranova, O. V., Hong, K., Sowers, L. C., & Zhang, Y. (2010). Role of Tet proteins in 5mC to 5hmC conversion, ES-cell self-renewal and inner cell mass specification. *Nature, 466*(7310), 1129–1133. http://dx.doi.org/10.1038/nature09303.

Jakovcevski, M., & Akbarian, S. (2012). Epigenetic mechanisms in neurological disease. *Nature Medicine, 18*(8), 1194–1204. http://dx.doi.org/10.1038/nm.2828.

Jin, S. G., Wu, X., Li, A. X., & Pfeifer, G. P. (2011). Genomic mapping of 5-hydroxymethylcytosine in the human brain. *Nucleic Acids Research, 39*(12), 5015–5024. http://dx.doi.org/10.1093/nar/gkr120.

Jones, P. A., & Baylin, S. B. (2002). The fundamental role of epigenetic events in cancer. *Nature Reviews Genetics, 3*(6), 415–428. http://dx.doi.org/10.1038/nrg816.

Kaas, G. A., Zhong, C., Eason, D. E., Ross, D. L., Vachhani, R. V., Ming, G. L., et al. (2013). TET1 controls CNS 5-methylcytosine hydroxylation, active DNA demethylation, gene transcription, and memory formation. *Neuron, 79*(6), 1086–1093. http://dx.doi.org/10.1016/j.neuron.2013.08.032.

Ko, M., Bandukwala, H. S., An, J., Lamperti, E. D., Thompson, E. C., Hastie, R., et al. (2011). Ten-eleven-translocation 2 (TET2) negatively regulates homeostasis and differentiation of hematopoietic stem cells in mice. *Proceedings of the National Academy of Sciences of the United States of America, 108*(35), 14566–14571. http://dx.doi.org/10.1073/pnas.1112317108.

Koh, K. P., Yabuuchi, A., Rao, S., Huang, Y., Cunniff, K., Nardone, J., et al. (2011). Tet1 and Tet2 regulate 5-hydroxymethylcytosine production and cell lineage specification in mouse embryonic stem cells. *Cell Stem Cell, 8*(2), 200–213. http://dx.doi.org/10.1016/j.stem.2011.01.008.

Kumar, R., DiMenna, L., Schrode, N., Liu, T. C., Franck, P., Munoz-Descalzo, S., et al. (2013). AID stabilizes stem-cell phenotype by removing epigenetic memory of pluripotency genes. *Nature, 500*(7460), 89–92. http://dx.doi.org/10.1038/nature12299.

Li, E. (2002). Chromatin modification and epigenetic reprogramming in mammalian development. *Nature Reviews Genetics, 3*(9), 662–673. http://dx.doi.org/10.1038/nrg887.

Li, E., Bestor, T. H., & Jaenisch, R. (1992). Targeted mutation of the DNA methyltransferase gene results in embryonic lethality. *Cell, 69*(6), 915–926.

Li, Z., Gu, T. P., Weber, A. R., Shen, J. Z., Li, B. Z., Xie, Z. G., et al. (2015). Gadd45a promotes DNA demethylation through TDG. *Nucleic Acids Research, 43*(8), 3986–3997. http://dx.doi.org/10.1093/nar/gkv283.

Li, X., Wei, W., Zhao, Q. Y., Widagdo, J., Baker-Andresen, D., Flavell, C. R., et al. (2014). Neocortical Tet3-mediated accumulation of 5-hydroxymethylcytosine promotes rapid behavioral adaptation. *Proceedings of the National Academy of Sciences of the United States of America, 111*(19), 7120–7125. http://dx.doi.org/10.1073/pnas.1318906111.

Li, E., & Zhang, Y. (2014). DNA methylation in mammals. *Cold Spring Harbor Perspectives in Biology, 6*(5), a019133. http://dx.doi.org/10.1101/cshperspect.a019133.

Lim, D. A., Huang, Y. C., Swigut, T., Mirick, A. L., Garcia-Verdugo, J. M., Wysocka, J., et al. (2009). Chromatin remodelling factor Mll1 is essential for neurogenesis from postnatal neural stem cells. *Nature, 458*(7237), 529–533. http://dx.doi.org/10.1038/nature07726 pii:nature07726.

Lopez-Serra, L., & Esteller, M. (2008). Proteins that bind methylated DNA and human cancer: reading the wrong words. *British Journal of Cancer, 98*(12), 1881–1885. http://dx.doi.org/10.1038/sj.bjc.6604374.

Lu, X., Song, C. X., Szulwach, K., Wang, Z., Weidenbacher, P., Jin, P., et al. (2013). Chemical modification-assisted bisulfite sequencing (CAB-Seq) for 5-carboxylcytosine detection in DNA. *Journal of the American Chemical Society, 135*(25), 9315–9317. http://dx.doi.org/10.1021/ja4044856.

Ma, D. K., Jang, M. H., Guo, J. U., Kitabatake, Y., Chang, M. L., Pow-Anpongkul, N., et al. (2009). Neuronal activity-induced Gadd45b promotes epigenetic DNA demethylation and adult neurogenesis. *Science, 323*(5917), 1074–1077. http://dx.doi.org/10.1126/science.1166859.

Ma, D. K., Marchetto, M. C., Guo, J. U., Ming, G. L., Gage, F. H., & Song, H. (2010). Epigenetic choreographers of neurogenesis in the adult mammalian brain. *Nature Neuroscience, 13*(11), 1338–1344. http://dx.doi.org/10.1038/nn.2672.

Maiti, A., & Drohat, A. C. (2011). Thymine DNA glycosylase can rapidly excise 5-formylcytosine and 5-carboxylcytosine: potential implications for active demethylation of CpG sites. *The Journal of Biological Chemistry, 286*(41), 35334–35338. http://dx.doi.org/10.1074/jbc.C111.284620.

Ming, G. L., & Song, H. (2011). Adult neurogenesis in the mammalian brain: significant answers and significant questions. *Neuron, 70*(4), 687–702. http://dx.doi.org/10.1016/j.neuron.2011.05.001 pii:S0896-6273(11)00348-5.

Morgan, H. D., Dean, W., Coker, H. A., Reik, W., & Petersen-Mahrt, S. K. (2004). Activation-induced cytidine deaminase deaminates 5-methylcytosine in DNA and is expressed in pluripotent tissues: implications for epigenetic reprogramming. *The Journal of Biological Chemistry, 279*(50), 52353–52360. http://dx.doi.org/10.1074/jbc.M407695200.

Nabel, C. S., Jia, H., Ye, Y., Shen, L., Goldschmidt, H. L., Stivers, J. T., et al. (2012). AID/APOBEC deaminases disfavor modified cytosines implicated in DNA demethylation. *Nature Chemical Biology, 8*(9), 751–758. http://dx.doi.org/10.1038/nchembio.1042.

Neri, F., Incarnato, D., Krepelova, A., Rapelli, S., Anselmi, F., Parlato, C., et al. (2015). Single-base resolution analysis of 5-formyl and 5-carboxyl cytosine reveals promoter DNA methylation dynamics. *Cell Reports.* http://dx.doi.org/10.1016/j.celrep.2015.01.008.

Nguyen, S., Meletis, K., Fu, D., Jhaveri, S., & Jaenisch, R. (2007). Ablation of de novo DNA methyltransfer-ase Dnmt3a in the nervous system leads to neuromuscular defects and shortened lifespan. *Developmental Dynamics: An Official Publication of the American Association of Anatomists, 236*(6), 1663–1676. http://dx.doi.org/10.1002/dvdy.21176.

Noctor, S. C., Martinez-Cerdeno, V., Ivic, L., & Kriegstein, A. R. (2004). Cortical neurons arise in symmetric and asymmetric division zones and migrate through specific phases. *Nature Neuroscience, 7*(2), 136–144. http://dx.doi.org/10.1038/nn1172 pii:nn1172.

Okano, M., Bell, D. W., Haber, D. A., & Li, E. (1999). DNA methyltransferases Dnmt3a and Dnmt3b are essential for de novo methylation and mammalian development. *Cell, 99*(3), 247–257.

Palmer, B. R., & Marinus, M. G. (1994). The dam and dcm strains of Escherichia coli–a review. *Gene, 143*(1), 1–12.

Pastor, W. A., Pape, U. J., Huang, Y., Henderson, H. R., Lister, R., Ko, M., et al. (2011). Genome-wide map-ping of 5-hydroxymethylcytosine in embryonic stem cells. *Nature, 473*(7347), 394–397. http://dx.doi.org/10.1038/nature10102.

Popp, C., Dean, W., Feng, S., Cokus, S. J., Andrews, S., Pellegrini, M., et al. (2010). Genome-wide erasure of DNA methylation in mouse primordial germ cells is affected by AID deficiency. *Nature, 463*(7284), 1101–1105. http://dx.doi.org/10.1038/nature08829.

Raiber, E. A., Beraldi, D., Ficz, G., Burgess, H. E., Branco, M. R., Murat, P., et al. (2012). Genome-wide distribution of 5-formylcytosine in embryonic stem cells is associated with transcription and depends on thymine DNA glycosylase. *Genome Biology, 13*(8), R69. http://dx.doi.org/10.1186/gb-2012-13-8-r69.

Raiber, E. A., Murat, P., Chirgadze, D. Y., Beraldi, D., Luisi, B. F., & Balasubramanian, S. (2015). 5-Formylcy-tosine alters the structure of the DNA double helix. *Nature Structural & Molecular Biology, 22*(1), 44–49. http://dx.doi.org/10.1038/nsmb.2936.

Reisenauer, A., Kahng, L. S., McCollum, S., & Shapiro, L. (1999). Bacterial DNA methylation: a cell cycle regulator? *Journal of Bacteriology, 181*(17), 5135–5139.

Riggs, A. D. (1975). X inactivation, differentiation, and DNA methylation. *Cytogenetics and Cell Genetics, 14*(1), 9–25.

Robertson, A. B., Dahl, J. A., Vagbo, C. B., Tripathi, P., Krokan, H. E., & Klungland, A. (2011). A novel method for the efficient and selective identification of 5-hydroxymethylcytosine in genomic DNA. *Nucleic Acids Research, 39*(8), e55. http://dx.doi.org/10.1093/nar/gkr051.

Rudenko, A., Dawlaty, M. M., Seo, J., Cheng, A. W., Meng, J., Le, T., et al. (2013). Tet1 is critical for neuronal activity-regulated gene expression and memory extinction. *Neuron, 79*(6), 1109–1122. http://dx.doi.org/10.1016/j.neuron.2013.08.003.

Shen, L., Wu, H., Diep, D., Yamaguchi, S., D'Alessio, A. C., Fung, H. L., et al. (2013). Genome-wide analysis reveals TET- and TDG-dependent 5-methylcytosine oxidation dynamics. *Cell, 153*(3), 692–706. http://dx.doi.org/10.1016/j.cell.2013.04.002.

Siegel, A., & Sapru, H. N. (2015). *Essential neuroscience* (3rd ed.). Baltimore, Philadelphia: Wolters Kluwer.

Song, C. X., Szulwach, K. E., Dai, Q., Fu, Y., Mao, S. Q., Lin, L., et al. (2013). Genome-wide profiling of 5-formylcytosine reveals its roles in epigenetic priming. *Cell, 153*(3), 678–691. http://dx.doi.org/10.1016/j.cell.2013.04.001.

Song, C. X., Szulwach, K. E., Fu, Y., Dai, Q., Yi, C., Li, X., et al. (2011). Selective chemical labeling reveals the genome-wide distribution of 5-hydroxymethylcytosine. *Nature Biotechnology, 29*(1), 68–72. http://dx.doi.org/10.1038/nbt.1732.

Stroud, H., Feng, S., Morey Kinney, S., Pradhan, S., & Jacobsen, S. E. (2011). 5-Hydroxymethylcytosine is associated with enhancers and gene bodies in human embryonic stem cells. *Genome Biology, 12*(6), R54. http://dx.doi.org/10.1186/gb-2011-12-6-r54.

Sun, Z., Dai, N., Borgaro, J. G., Quimby, A., Sun, D., Correa, I. R., Jr., et al. (2015). A sensitive approach to map genome-wide 5-hydroxymethylcytosine and 5-formylcytosine at single-base resolution. *Molecular Cell, 57*(4), 750–761. http://dx.doi.org/10.1016/j.molcel.2014.12.035.

Suzuki, M. M., & Bird, A. (2008). DNA methylation landscapes: provocative insights from epigenomics. *Nature Reviews Genetics, 9*(6), 465–476. http://dx.doi.org/10.1038/nrg2341.

Tahiliani, M., Koh, K. P., Shen, Y., Pastor, W. A., Bandukwala, H., Brudno, Y., et al. (2009). Conversion of 5-methylcytosine to 5-hydroxymethylcytosine in mammalian DNA by MLL partner TET1. *Science, 324*(5929), 930–935. http://dx.doi.org/10.1126/science.1170116.

Terragni, J., Bitinaite, J., Zheng, Y., & Pradhan, S. (2012). Biochemical characterization of recombinant beta-glucosyltransferase and analysis of global 5-hydroxymethylcytosine in unique genomes. *Biochemistry*, *51*(5), 1009–1019. http://dx.doi.org/10.1021/bi2014739.

Urdinguio, R. G., Sanchez-Mut, J. V., & Esteller, M. (2009). Epigenetic mechanisms in neurological diseases: genes, syndromes, and therapies. *Lancet Neurology*, *8*(11), 1056–1072. http://dx.doi.org/10.1016/s1474-4422(09)70262-5.

Wang, L., Zhou, Y., Xu, L., Xiao, R., Lu, X., Chen, L., et al. (2015). Molecular basis for 5-carboxycytosine recognition by RNA polymerase II elongation complex. *Nature*, *523*(7562), 621–625. http://dx.doi.org/10.1038/nature14482.

Wheldon, L. M., Abakir, A., Ferjentsik, Z., Dudnakova, T., Strohbuecker, S., Christie, D., et al. (2014). Transient accumulation of 5-carboxylcytosine indicates involvement of active demethylation in lineage specification of neural stem cells. *Cell Reports*, *7*(5), 1353–1361. http://dx.doi.org/10.1016/j.celrep.2014.05.003.

Williams, K., Christensen, J., Pedersen, M. T., Johansen, J. V., Cloos, P. A., Rappsilber, J., et al. (2011). TET1 and hydroxymethylcytosine in transcription and DNA methylation fidelity. *Nature*, *473*(7347), 343–348. http://dx.doi.org/10.1038/nature10066.

Wu, H., D'Alessio, A. C., Ito, S., Wang, Z., Cui, K., Zhao, K., et al. (2011). Genome-wide analysis of 5-hydroxymethylcytosine distribution reveals its dual function in transcriptional regulation in mouse embryonic stem cells. *Genes & Development*, *25*(7), 679–684. http://dx.doi.org/10.1101/gad.2036011.

Xu, Y., Wu, F., Tan, L., Kong, L., Xiong, L., Deng, J., et al. (2011). Genome-wide regulation of 5hmC, 5mC, and gene expression by Tet1 hydroxylase in mouse embryonic stem cells. *Molecular Cell*, *42*(4), 451–464. http://dx.doi.org/10.1016/j.molcel.2011.04.005.

Yao, B., & Jin, P. (2014). Unlocking epigenetic codes in neurogenesis. *Genes & Development*, *28*(12), 1253–1271. http://dx.doi.org/10.1101/gad.241547.114 pii:28/12/1253.

Yu, H., Su, Y., Shin, J., Zhong, C., Guo, J. U., Weng, Y. L., et al. (2015). Tet3 regulates synaptic transmission and homeostatic plasticity via DNA oxidation and repair. *Nature Neuroscience*, *18*(6), 836–843. http://dx.doi.org/10.1038/nn.4008.

Zhang, R. R., Cui, Q. Y., Murai, K., Lim, Y. C., Smith, Z. D., Jin, S., et al. (2013). Tet1 regulates adult hippocampal neurogenesis and cognition. *Cell Stem Cell*, *13*(2), 237–245. http://dx.doi.org/10.1016/j.stem.2013.05.006.

Zhao, C., Deng, W., & Gage, F. H. (2008). Mechanisms and functional implications of adult neurogenesis. *Cell*, *132*(4), 645–660. http://dx.doi.org/10.1016/j.cell.2008.01.033. pii:S0092-8674(08)00134-7.

CHAPTER 4

TET and 5hmC in Neurodevelopment and the Adult Brain

M. Fasolino, S.A. Welsh, Z. Zhou
University of Pennsylvania, Philadelphia, PA, United States

INTRODUCTION

DNA methylation at the 5-carbon of cytosine (C) is widely distributed throughout the mammalian genome, with 5% of all C and 85% of all cytosine-phosphate-guanine dinucleotides (CpGs) being methylated (Lister et al., 2013). Such methylation plays an essential role in various biological functions such as the regulation of gene transcription, establishment and maintenance of cellular identity, imprinting, silencing of transposons and repetitive elements, and chromosome X inactivation (Jaenisch & Bird, 2003). Historically, DNA methylation of C was thought to be a stable covalent modification, existing exclusively as 5-methylcytosine (5mC). However, this view was challenged in 2009 when two seminal papers published in parallel described another C modification, 5-hydroxymethylcytosine (5hmC), which is formed from the oxidation of 5mC (Kriaucionis & Heintz, 2009; Tahiliani et al., 2009). Tahiliani et al. (2009) also described enzymes that were able to convert 5mC to 5hmC, the Ten-eleven translocation family of enzymes, or Tet enzymes. These enzymes were found to be paralogues of the base J binding proteins (JBPs) from the parasite *Trypanosoma brucei*. However, instead of converting the base thymine to 5-hydroxymethyl-uracil, as JBP enzymes do, Tet enzymes convert 5mC to 5hmC.

Since its rediscovery in 2009, 5hmC has added an important dimension in understanding the epigenetic regulation of neuronal function (Kriaucionis & Heintz, 2009; Penn, Suwalski, O'Riley, Bojanowski, & Yura, 1972; Wyatt & Cohen, 1953). The particular importance of 5hmC in the brain is highlighted by the fact that although global 5mC levels are similar across different tissue types, levels of 5hmC are highly variable, with the highest concentration in the central nervous system (CNS) (Globisch et al., 2010; Kriaucionis & Heintz, 2009; Münzel et al., 2010). Notably, all mature neurons in the CNS are postmitotic, meaning that they no longer divide. Although it was previously known that 5mC could be passively removed through cell division, the discovery of Tet enzymes meant that 5mC could be actively removed via oxidation by Tets to 5hmC, and this removal could occur in postmitotic cells. Furthermore, 5hmC can also be removed, completely reverting the base back to unmodified C. Removal of 5hmC occurs first via

DNA Modifications in the Brain
ISBN 978-0-12-801596-4
http://dx.doi.org/10.1016/B978-0-12-801596-4.00004-6

61

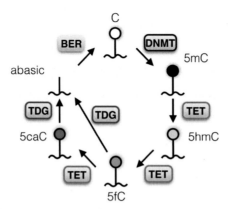

Figure 4.1 *Cytosine modification cycle.* Unmodified cytosine (C) is converted to 5-methylcytosine (5mC) by DNA methyltransferase (DNMT) enzymes DNMT1, DNMT3a, or DNMT3b. 5mC can then be iteratively oxidized by Ten-eleven translocation (Tet) enzymes Tet1, Tet2, or Tet3 to become 5-hydroxymethylcytosine (5hmC), 5-formylcytosine (5fC), and 5-carboxylcytosine (5caC). Thymine-DNA glycosylase (TDG) can recognize the bases of 5fC and 5caC and excise them from the DNA, leaving an abasic site. An abasic site triggers the base excision repair pathway (BER), which restores the base to C.

iterative oxidation by the Ten-eleven translocation (TET) family of proteins (TET1, TET2, and TET3, collectively referred to as TETs), which convert 5hmC to 5-formyl-cytosine (5fC), and subsequently to 5-carboxylcytosine (5caC) (Ito et al., 2011). Finally, 5caC is converted to C by thymine-DNA glycosylase (TDG)–mediated base excision repair (BER) (He et al., 2011) (Fig. 4.1). Therefore, it has been called into question whether 5hmC is a mere transient, uninformative by-product of DNA demethylation or a stable, purposeful epigenetic mark with biological functional significance. Throughout this chapter, we highlight recent findings that have greatly advanced our understanding of the role of 5hmC in brain.

GLOBAL 5hmC AND TET EXPRESSION THROUGHOUT THE ADULT BRAIN

To gain a better understanding of the biological function of 5hmC, initial studies interrogated the levels of 5hmC in various tissues by using isotope-based liquid chromatography-mass spectrometry (LC-MS). It has been unanimously found that although this mark is present throughout the body, 5hmC levels are markedly higher in the nervous system than other tissues. 5hmC concentrations in the CNS are brain region-specific: 5hmC constitutes 0.7% of total C bases in the cortex and hypothalamus; 0.6% in the brainstem, olfactory bulb, and hippocampus; 0.5% in the spinal cord; 0.4% in the mid-brain; and 0.3% in the cerebellum (Globisch et al., 2010; Kriaucionis & Heintz, 2009; Münzel et al., 2010). For comparison, tissues with the next closest levels of 5hmC, ranging from 0.15% to 0.17%, are the kidney, nasal epithelium, bladder, heart, muscle, and lung. In the pituitary gland, a non-neuronal structure at the base of the brain, 5hmC constitutes only 0.06% of total C, which supports that high concentrations of 5hmC are specific to neural tissues, rather than topographic location (Globisch et al., 2010). Further support of high levels of 5hmC being specific to neurons comes from a comprehensive genome-wide study that found 5hmC levels are higher in neuronal than in non-neuronal cell types within the frontal cortex (Lister et al., 2013).

It remains unknown what accounts for the high levels of global 5hmC within the brain as compared to other tissues. Since Tets are expressed at similar levels in other tissues with much lower 5hmC levels, overall expression levels of this family of proteins cannot account for the difference. It also remains to be determined what accounts for the differences in 5hmC levels across brain regions. Although the cortex, hippocampus, and cerebellum have vastly different 5hmC levels that range from 0.7% to 0.3% of total C, the regions have similar Tet expression levels (Szwagierczak, Bultmann, Schmidt, Spada, & Leonhardt, 2010). It is possible that brain region-specific levels of 5hmC are dependent upon the relative levels of TET-dependent cofactors, iron and α-ketoglutarate (α-KG) (Tahiliani et al., 2009), or other potential regulators, such as ascorbic acid, calpain, succinate, and fumarate. In embryonic stem cell (ESC) culture, supplementation of ascorbic acid decreased the levels of 5mC and increased levels of 5hmC (Blaschke et al., 2013; Wang & Zhang, 2014; Yin et al., 2013). These effects have unique implications for TET activity in the brain, since the brain has the highest levels of ascorbic acid in the body, mostly due to high expression of the ascorbic acid–specific transporter SVCT2 (Yin et al., 2013). More research is this area is warranted to elucidate what controls the levels of 5hmC.

In addition to brain region specificity, global 5hmC levels are also cell type specific and developmentally dependent. The cell type-specificity of 5hmC was noted in the paper that rediscovered it in 2009 with the description that 5hmC constitutes 0.6% of total CpGs in Purkinje neurons, but only 0.2% of granule cells (Kriaucionis & Heintz, 2009). This discovery has been confirmed by an independent group finding that the levels and genomic distribution of 5hmC vary across Purkinje cells, granule cells, and Bergmann glia of the cerebellum (Mellén, Ayata, Dewell, Kriaucionis, & Heintz, 2012). The establishment of these cell type differences in 5hmC levels in the brain occurs between embryonic days 12.5 and 13.5, which is the time at which active specification of neurons and glial cells commences, supporting that the 5hmC is important for neuronal identity (Wheldon et al., 2014). The rapid increase in global 5hmC during neuronal differentiation and synaptogenesis also highlights the developmental importance of this epigenetic mark in the brain (Hahn et al., 2013; Song et al., 2011; Szulwach et al., 2011; Lister et al., 2013).

In the adult mouse (10–12 week) brain, neuronal activity can also lead to changes in global 5hmC. When neuronal activity is increased via flurothyl-induced seizures, there is a significant reduction in 5mC and 5hmC in cornu ammonis 1 (CA1) of the hippocampus of adult mice at 24 h after the occurrence of seizures (Kaas et al., 2013). This effect differs from that of the dentate granule (DG) of the hippocampus cells in that synchronous activation of DG cells in the adult mice by electroconvulsive stimulation or voluntary running does not lead to a global change in methylation, as there are a similar number of CpGs that become methylated and demethylated under these conditions as assessed by methyl-sensitive cut counting method (Guo et al., 2011a).

These findings on the global 5hmC patterns throughout the brain and other tissues suggest that 5hmC might be especially important in neuronal function and identity. However, to truly unravel the biology of 5hmC, determining the genomic location of this epigenetic mark and its effect on gene expression is necessary, and is discussed in the following section.

GENOMIC DISTRIBUTION OF 5hmC

Recent advances in sequencing have allowed for the mapping of the genomic distribution of 5hmC in the brain, providing invaluable insight into the biological function of this epigenetic mark. Various affinity- and enzyme-based methods have been developed for profiling 5hmC genome wide, with three most commonly used approaches. First is 5hmC selective chemical labeling in which 5hmC is converted to biotin-N3-5-hydroxymethyl-cytosine for affinity enrichment through a two step synthesis (Song et al., 2011). Second is hydroxymethylated DNA immunoprecipitation in which 5hmC is enriched via antibodies that specifically bind to 5hmC (Jin, Wu, Li, & Pfeifer, 2011). Third is TET-assisted bisulfite sequencing (TAB-seq) in which 5hmC is exclusively protected via glycosylation and TET-mediated oxidation before bisulfite treatment (Yu et al., 2012).

With use of these approaches, general features have emerged. Quantitatively, intragenic and global 5hmC levels are equivalent across chromosomes in both human and mouse, except for the male chromosome X, which has 22% lower enrichment (Lister et al., 2013; Mellén et al., 2012; Szulwach et al., 2011). 5hmC is predominantly found in CpGs in both human and mouse across development (Lister et al., 2013; Wen et al., 2014). In the fetal mouse brain, 5% of CpGs and 0% of non-CpGs (CpH, where H=A, C, or T) are hydroxymethylated, whereas in the adult mouse frontal cortex (6 week), hydroxymethylation occurs at 19% of CpGs and 0.02% of CHs. This epigenetic mark mostly is found across transcriptional end sites, intragenic regions, DNase I-hypersensitive sites (DHSs), and enhancers (Lister et al., 2013). It is present at both poised enhancers (solely marked by mono-methylated Lysine 4 of histone H3 (H3K4me1)) and active enhancers (marked by both H3K4me1 and acetylated Lysine 27 of histone H3 (H3K27ac)). Major satellite and promoter regions, in contrast, are relatively devoid of 5hmC (Wen et al., 2014). Most 5hmC (71%) is found intragenically, with a much higher concentration at exons than introns (Szulwach et al., 2011). These findings on the genomic distribution of 5hmC implicate that it may play a role in gene regulation.

Given the relatively high enrichment of 5hmC across exons, and the proposed hypothesis that methylation modulates alternative splicing (Maunakea, Chepelev, Cui, & Zhao, 2013), studies have evaluated the role of this epigenetic mark in splicing. 5hmC seems to play an important role in alternative exon use in the mammalian brain, as there is a distinct pattern of methylation at exon–intron boundaries. First, there is a sharp decrease in 5hmC at the 5′ end of the intron at the exon–intron boundary. Second, across exons from 5′ to 3′, there is a substantial increase in 5mC levels and a less pronounced decrease in 5hmC (Khare et al., 2012; Wen et al., 2014). Third, 5hmC levels, but not 5mC levels, within 20 bp of the exon–intron boundary correlate with constitutively used exons relative to alternatively spliced exons. The importance of these features in alternative exon usage rather than general transcription is highlighted by the fact that first exons have much lower 5mC and 5hmC than internal exons and that exons of intron-less or single-exon genes have lower 5hmC than multiple-exon genes (Khare et al., 2012). This feature seems to be specific to brain tissue since neither 5mC nor 5hmCs correlate with exon use in the liver. Third, flanking the highly conserved "GT" splice site sequence at the 5′ splicing sites (5′ ss) of internal exons, at the −1 and −2 positions on the exon side and +4 and +5 positions of the intron side of the exon–intron boundary, are two prominent 5hmC peaks. 5mC, in contrast, does not exhibit this type of pattern in the brain (Khare et al., 2012; Wen et al., 2014). This patterning of 5hmC at the 5′ ss seems to be brain specific as 5mC, rather than 5hmC, marks exon–intron boundaries in the liver (Khare et al., 2012). Further examination of alternatively spliced exons by RNA-seq found that low or no methylation flanking the 5′ ss is associated with significantly more exon skipping than methylated or hydroxymethylated boundaries. This suggests that demethylation is associated with alternative splicing events, which is consistent with the idea that 5hmC aids in exon recognition and inclusion (Khare et al., 2012; Maunakea et al., 2013; Wen et al., 2014).

In addition to the correlation of 5hmC in exon use, there is also a strong positive correlation between intragenic 5hmC levels and gene expression in both main cell types of the brain, neurons and glia (Lister et al., 2013; Mellén et al., 2012; Song et al., 2011). 5mC levels across the gene body, in contrast, negatively correlate with gene expression (Lister et al., 2013; Mellén et al., 2012; Wen et al., 2014). The best correlate with gene expression is the intragenic ratio of 5hmC to 5mC (5hmC/5mC). This correlation extends to the tissue-specific and cell subtype–specific level, with a relatively high 5hmC/5mC ratio correlating with brain region–specific and cell type–specific differentially expressed transcripts (Lister et al., 2013; Mellén et al., 2012). When the 5hmC genomic distribution and expression profiles of three different cell types of the cerebellum (Purkinje cells, granule cells, and Bergmann glia) were compared, it was found that cell type–specific transcripts have higher intragenic 5hmC/5mC levels than the other cell types (Mellén et al., 2012). This cell type–specific patterning also holds true across neuronal differentiation as cell type–specific genes that are developmentally regulated gain intragenic 5hmC and lose intragenic 5mC across differentiation (Colquitt, Allen, Barnea, & Lomvardas, 2013).

There is also a significant difference between 5hmC and 5mC for strand bias of expressed genes in both glia and neurons. When comparing the lowest to the highest expressed genes, there is a seven-fold bias in 5hmC enrichment on the sense strand and a five-fold bias in 5mC enrichment on the antisense strand. These findings suggest that 5hmC enrichment on the sense strand is correlated with activation. In agreement with this is the finding that 5hmC is inversely related to two repressive histone modifications, tri-methylated Lysine 27 of histone H3 (H3K27me3) and tri-methylated Lysine 9 of histone H3 (H3K9me3). Alternatively, these two histone modifications correlate with 5mC (Wen et al., 2014). Genes enriched for 5hmC in the mammalian brain relative to other tissues are synapse related (Khare et al., 2012). These findings highlight the importance of 5hmC in the activation of genes specific to the brain, neuronal subtypes, and neuronal function.

Although 5hmC levels correlate highly with gene expression, it is unknown how this correlation is accomplished. What are the mechanisms that allow 5hmC to influence or be influenced by gene expression? It is hypothesized that there are specific proteins that are able to bind 5hmC and influence gene expression through binding to additional protein complexes or by initiating a signaling cascade(s). The first paper to inquire into what proteins can bind 5hmC used quantitative MS-based proteomics. They used fragments of DNA that contained 5hmC to isolate interacting proteins from mouse ESCs and analyzed the resulting proteins by using LC-MS/MS. They identified a large number of potential 5hmC-interacting proteins, most notably N-methylpurine-DNA glycosylase and nei like 3, DNA glycosylases that, like TDG, may participate in the active DNA demethylation pathway to convert 5hmC to unmodified C via BER. Interestingly, they found proteins that had previously been uncharacterized, such as WD repeat domain 76 (Wdr76). By purifying Wdr76 they identified Wdr76-interacting proteins, including a DNA helicase, Hells (a lymphoid specific helicase), that is thought to regulate DNA methylation levels and a protein that binds tri-methylated Lysine 4 of histone H3 (H3K4me3), Spindlin 1 (Spin 1). Looking at adult mouse brain cells, they confirmed the interaction of 5hmC with Wdr76 and thymocyte nuclear protein 1 (Thy28). A brain-specific 5hmC interaction was found with THAP domain containing 11, which is highly expressed in Purkinje cells. Additionally, they found that the proteins Thy28, ubiquitin-like with PHD and Ring finger domains (Uhrf1), and methyl-CpG-binding protein 2 (MeCP2) bind to both 5mC and 5hmC, although MeCP2 binds to 5mC with much higher affinity. The authors concluded that 5hmC is an active intermediate in DNA demethylation and may be involved in global epigenetic regulation (Spruijt et al., 2013).

However, the binding of MeCP2 to 5hmC is a contentious finding. Previous in vitro studies found that conversion of 5mC to 5hmC abolished binding of MeCP2 to oligonucleotide sequences (Valinluck et al., 2004). Another study compared the affinity of Uhrf1 and MeCP2 to modified DNA in vitro, and found that Uhrf1 had a similar affinity for 5mC and 5hmC, whereas MeCP2 had a greater affinity for 5mC, as shown previously (Frauer et al., 2011; Spruijt et al., 2013). Similarly, yet another independent group found that although MeCP2 is able to bind 5hmC, the affinity is 20-fold less than that of 5mC (Khrapunov et al., 2014). In contrast, it has been reported that MeCP2 binds 5mC and 5hmC with similar affinity and that a Rett-associated mutation in MeCP2 causes the disruption of its binding preferentially to 5hmC in vitro (Mellén et al., 2012). Additionally, Baubec, Ivanek, Lienert, & Schübeler (2013) found that MeCP2 localization correlates with 5hmC in ESCs. Further confounding results come from two studies addressing the affects of MeCP2 on levels of 5hmC in vivo (Mellén et al., 2012; Szulwach et al., 2011). Szulwach et al. (2011) showed that decreased levels of MeCP2 correlated with higher levels of 5hmC and that overexpression of MeCP2 revealed a decrease in 5hmC in the cerebellum. In contrast, Mellén et al. (2012) reported that loss of MeCP2 results in a small, but significant, decrease in 5hmC levels. Together, these findings suggest that further research is warranted to determine whether MeCP2 is a bona fide binding partner of 5hmC in vivo.

ROLE OF 5hmC IN BRAIN DEVELOPMENT

Across brain development in both human and mouse, 5hmC levels increase, with the adult human prefrontal cortex containing 10-fold more 5hmC than the fetal brain (Lister et al., 2013; Wen et al., 2014). In the adult (6-week-old mouse), high levels of 5hmC are found across enhancer, transcriptional end site, intragenic, and DHS regions. In the fetus, in contrast, enrichment of 5hmC is primarily in DHS regions and enhancer regions that are unique to fetal development (Lister et al., 2013). Many of the gene bodies that gain 5hmC enrichment in the adult stage also have 5hmC present at the fetal stage, albeit at much lower levels. This implies that the cell type–specific increase in 5hmC observed over development occurs at intragenic regions that were partially established at the fetal stage. Although 5hmC levels of the fetal brain are much lower than those of the adult brain across all genomic features, there are numerous CpGs in which hydroxymethylation is higher in the fetal brain (fetal > adult hydroxymethylated CpGs). These fetal > adult hydroxymethylated CpGs are enriched at enhancers (Lister et al., 2013; Wen et al., 2014). Developmentally downregulated genes have high levels of gene body 5hmC at the fetal stage, but not at the adult stage (Lister et al., 2013). Regions that have significantly higher hydroxymethylation in the fetal frontal cortex than the adult are dormant regions poised for demethylation and activation over development in both the mouse and human (Lister et al., 2013; Wen et al., 2014). Analysis of these developmentally dependent, differentially hydroxymethylated regions in adult Tet2−/− mice revealed that these 5hmC-poised loci are dependent upon TET2 activity (Lister et al., 2013). These findings suggest a key role of 5hmC in brain development that is conserved across mammals.

5hmC marks both developmentally dependent [postnatal day 7 (P7), 6-week-old, and 1-year-old mice] and brain region–specific (hippocampus compared to cerebellum) loci. When brain region–specific differentially hydroxymethylated loci were evaluated across development, it was found that 5hmC marks region-specific genes during early development (at P7 or earlier) to facilitate in region-specific transcriptional programs. These tissue-specific differentially hydroxymethylated regions are enriched for a 21-nucleotide motif that might be critical for regulating specific gene expression programs. 5hmC-regulated regions across development within mouse cerebellum and hippocampus revealed that some loci are stable (5hmC acquired during P7 and maintained through 1 year of age), whereas others are dynamic (5hmC is not present at all time points). When repeat elements were assessed across development, it was found that, in both mouse hippocampus and cerebellum, 5hmC enrichment increases at short interspersed nuclear elements and long tandem repeats and decreases in long interspersed nuclear elements and satellites. In P7 cerebellum, but not adult, 5hmC is associated with the transcriptional start site (TSS) of genes with low expression. This can be explained by the fact that at this stage, a significant amount of progenitor cells are present, which is in line with the finding in mouse and human ESCs in which 5hmC is enriched at the TSS of repressed, but developmentally poised, genes. At developmentally activated genes, there is an increase in intragenic 5hmC, whereas at developmentally repressed genes, there is only a small decrease in 5hmC across the gene body (Szulwach et al., 2011). These findings highlight the importance of 5hmC in brain development.

5hmC CHANGES ASSOCIATED WITH NEURONAL DIFFERENTIATION

In addition to developmentally dependent changes in 5hmC, alterations in this epigenetic mark also occur across neuronal differentiation as assessed by comparing the genomic distribution of 5hmC in neural progenitor cells (NPCs) in the subventricular zone to that of maturing neurons of the cortical plate in embryonic day 15.5 mice. In NPCs and maturing neurons alike, there is an absence of 5hmC at enhancers and enrichment at promoters and gene bodies. Over differentiation there is an increase in 5hmC (but not 5mC), which is primarily intragenic. As typically found, intragenic 5hmC correlates transcription in both progenitors and mature neurons; however, this association is more pronounced in maturing neurons. Genes that gain intragenic 5hmC are associated with neuronal differentiation and axonogenesis. Genes with the highest increase in 5hmC over differentiation did not show an increase in demethylation, indicating the stability of this epigenetic mark. Concomitant with the increase in intragenic 5hmC during neuronal differentiation at activated genes is a gain of H3K4me3 at the promoter and a loss of H3K27me3 in gene bodies and promoters. TET2 or TET3 may play a role in neurogenesis since knockdown of their expression via electroporation of small hairpin RNAs (shRNAs) lead to defects in the progression of differentiation (Hahn et al., 2013).

Similar changes in 5hmC over neuronal differentiation also occurs in the olfactory sensory neuron (OSN) differentiation as there is an increase in 5hmC from the horizontal basal cell stage to the globose basal cell to mature olfactory sensory neurons (mOSNs).

Similar to neuronal differentiation in the ventricular zone, the increase in 5hmC across OSN differentiation is primarily intragenic and associated with developmentally regulated genes. When Tet3 is overexpressed in mOSNs via a transgenic mouse model approach, genes with modest levels of 5hmC in the wild-type (WT) mice exhibit an increase in 5hmC levels, and subsequent increase in expression, whereas genes with high levels of 5hmC exhibit a loss of 5hmC, and subsequent decrease in expression. The downregulation of the most highly expressed mOSN-specific genes, such as olfactory receptors and guidance molecules, affects glomerular formation (Colquitt et al., 2013). These findings from two neurogenic regions of the brain support the role of 5hmC in neuronal differentiation.

ROLE OF 5hmC IN AGING AND NEURODEGENERATION

Considering the importance of 5hmC in neurodevelopment, researchers have investigated the role of 5hmC in aging and neurodegenerative diseases. The two studies conducted on aging (from 6 weeks to 2 years) have found that global 5hmC levels increase in the hippocampus as aging occurs (Chen, Dzitoyeva, & Manev, 2012; Chouliaras et al., 2012). Given this finding, more research is warranted to determine whether 5hmC also increases in other brain regions, as well as the specific loci with 5hmC enrichment across aging. Although there are not an expansive number of studies on neurodegeneration, several correlations have been found with 5hmC for both Huntington's disease (HD) and Alzheimer's disease (AD). In a mouse model of HD, a genome-wide decrease in 5hmC levels was found, correlating with decreases in gene expression (Wang et al., 2013). The differentially hydroxymethylated regions contained genes involved in neuronal development and survival, which could have an important effect on the neurodegeneration seen in this disease (Wang et al., 2013). Additionally, studies in AD patients have seen a decrease in 5hmC in the hippocampus, and 5hmC levels were negatively correlated with amyloid plaque load, a common marker of AD pathogenesis (Chouliaras et al., 2013; Condliffe et al., 2014). However, another study looking at preclinical stages of AD saw an increase in 5hmC in hippocampal regions of the brain (Coppieters et al., 2014; Sun, Zang, Shu, & Li, 2014). Although preliminary, these studies point to a potential role of 5hmC in neurodegeneration through deregulation of gene expression. Additionally, 5hmC may act as a biomarker for diagnosing and determining the stage of the disease. Further research is required to understand the role 5hmC plays in neurodegeneration.

ROLE OF TET ENZYMES IN BRAIN FUNCTION

The three TET proteins, TET1, TET2, and TET3, are members of the Fe(II)/α-KG–dependent dioxygenase family of enzymes. It has been proposed that the enzymes use molecular oxygen to catalyze oxidative decarboxylation of α-KG, creating a highly reactive intermediate that converts 5mC to 5hmC (Fig. 4.2). This proposed mechanism is based off of other Fe(II)/α-KG–dependent dioxygenase family proteins, since a structure of mammalian TET enzymes has not been solved. Each of the TETs contain a conserved catalytic domain (double-stranded β-helix, DSBH, fold) that contains the metal-binding residues required for the oxidation reaction (Fig. 4.3) (Kohli & Zhang, 2013). Additionally, a cysteine-rich (Cys-rich) domain is found in all TET proteins upstream of the catalytic domain and is thought to be required for activity (Iyer, Tahiliani, Rao, & Aravind, 2009; Tahiliani et al., 2009). TET and TET3 contain a CXXC domain near the N-terminal end of the protein, which is known to bind to CpG sites (Kohli & Zhang, 2013). Although each of the human TET enzymes only share ~18–24% sequence identity (UniProt Consortium, 2015), the catalytic and Cys-rich domains are highly conserved between the three enzymes (Fig. 4.4). It is hypothesized that the remaining nonconserved portions of the protein may serve as regulatory domains and convey different functionality between the three TETs.

Tet1, Tet2, and Tet3 are all expressed in the brain, with Tet3 having the highest expression, followed by Tet2; Tet1 has much lower expression levels than the other two family

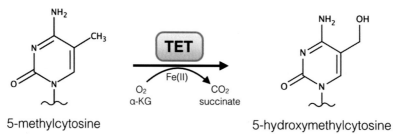

5-methylcytosine   5-hydroxymethylcytosine

Figure 4.2 *5-methylcytosine to 5-hydroxymethylcytosine conversion by Ten-eleven translocation enzymes.* Ten-eleven translocation (Tet) enzymes oxidize the 5-methyl group of 5-methylcytosine (5mC). With cofactors α-ketoglutarate (α-KG) and molecular oxygen, Tet oxidizes the 5-methyl carbon, adding a hydroxyl group, thereby yielding 5-hydroxymethylcytosine (5hmC). Other by-products of the enzymatic reaction include CO_2 and succinate.

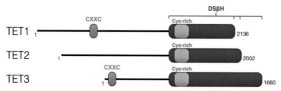

Figure 4.3 *Schematic of Ten-eleven translocation enzymes.* Ten-eleven translocation (Tet)1, Tet2, and Tet3 all share a conserved catalytic double-stranded β-helix fold (DSβH) domain at the C-terminal end of the protein. Additionally, a conserved cysteine-rich (Cys-rich) domain is found at the N-terminal portion of the catalytic domain. Tet1 and Tet3 contain an additional CpG-binding CXXC domain near the N-terminal end of the protein.

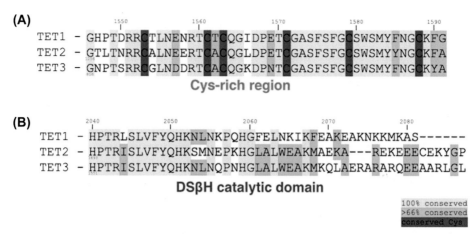

Figure 4.4 *Sequence conservation of Ten-eleven translocation domains.* (A) The cysteine-rich (Cys-rich) domain is almost fully conserved between the Ten-eleven translocation (Tet) enzymes. (B) The catalytic double-stranded β-helix fold (DSβH) domain is also highly conserved between the three Tet enzymes, even though the proteins only share 18–24% sequence identity. The remaining nonconserved domains are hypothesized to serve as regulatory domains.

members (Szwagierczak et al., 2010). Knockout, loss of function, and overexpression studies have revealed diverse functions of these enzymes, and the importance of 5hmC, in neuronal function.

TET1

Currently, Tet1 is the most well-studied Tet family member in the brain, most likely because TET1 was the first enzyme discovered to convert 5mC to 5hmC (Tahiliani et al., 2009). Although Tet1 expression is markedly lower than Tet2 and Tet3 in the brain, various studies have demonstrated the importance of Tet1 in neuronal function.

Tet1 whole-body knockout (KO) mice are viable and fertile without apparent health deficits, albeit a smaller body weight and litter size than WT animals (Dawlaty et al., 2011; Rudenko et al., 2013; Zhang et al., 2013). Additionally, there are no obvious morphological or developmental brain abnormalities (Rudenko et al., 2013; Zhang et al., 2013). In agreement with the importance of Tet1 in the generation of 5hmC, there is a small, but significant, reduction of 5hmC in the brains of Tet1 KO mice, but no change in 5mC. The fact that the change is small is likely due to presence of Tet2 and Tet3, the other members of the Tet family that are endogenously expressed at much higher levels than Tet1 in the brain. There are also no apparent deficits in synaptic connectivity as measured by Synapsin I, a marker of synaptic abundance (Rudenko et al., 2013). This compensation effect is supported by the abnormalities observed in Tet1 and Tet2 double knockout (DKO) mice. The majority of DKO die perinatally, although a small percentage survive without gross abnormalities. Compared to WT mice, DKO adult mice (2.5 months) have reduced 5hmC levels (34%) and increased 5mC levels (~5%) in the cerebrum and cerebellum. Although these are appreciable changes in methylation, a large portion of 5hmC remains intact, suggesting that Tet3 plays a critical role in its maintenance (Dawlaty et al., 2013).

Behaviorally, adult Tet1 single KO mice (4 months) do not show deficits in locomotion, anxiety, fear memory acquisition, or depression-related behaviors. Multiple groups have observed memory deficits; however, there is not a consensus as to the specific type of memory deficit. According to one group, Tet1 KO mice have impairments in short-term memory and spatial learning, but normal long-term memory, as assessed by Morris water maze (MWM) (Zhang et al., 2013). Another group reported normal short-term memory and spatial learning, but impaired spatial memory extinction in the MWM and classical Pavlovian fear conditioning (Rudenko et al., 2013). When Tet1 is overexpressed in the CA1 region of the hippocampus, long-term memory was affected (fear conditioning), but not locomotion, anxiety, or short-term memory. This deficit in long-term memory formation was observed for both catalytically active and inactive forms of TET1, suggesting that TET1's role in memory formation is independent of its catalytic activity. Tet1 expression, but not that of Tet2, Tet3, or other proteins involved in the demethylation pathway, is significantly downregulated in the dorsal CA1 of mice after fear learning (Kaas et al., 2013). These findings support that Tet1 contributes to basal neuronal 5hmC levels that are potentially important for neuronal function. The behavioral effects of Tet1 in the brain still warrant further investigation considering the confounding results.

At the cellular and molecular level, evidence suggests that Tet1 is important in neurogenesis and hippocampal function. When Tet1 KO mice were bred with Nestin-GFP transgenic mice, the number of GFP-positive cells in the subgranular zone in adult mice was dramatically reduced by 45% compared to WT animals (Zhang et al., 2013). This is different from two other non-neurogenic brain regions examined, the cingulate cortex and hippocampus CA1 (Rudenko et al., 2013). The reduction in proliferation potential of NPCs is likely to underlie this deficit as evidenced by a reduction in neurospheres isolated from Tet1 KO mice, the decrease in bromodeoxyuridine (BrdU, which marks dividing cells)-positive neurons in Tet1-dentate gyrus (DG)-knockdown in adult mice, and the 35% decrease in BrdU-positive neurons in animals in which Tet1 is specifically deleted in neural progenitors at 2 months of age. Examination of the gene expression and methylation changes in Tet1 KO mice revealed that the decreased expression of a cohort of genes involved in neurogenesis was associated with an increase in 5mC at their promoters, suggesting that TET1 positively regulates adult neurogenesis through the oxidation of 5mC to 5hmC at these genes (Zhang et al., 2013).

Tet1 overexpression in the DG or CA1 region of the hippocampus of mice results in a dramatic increase in 5hmC and decrease in 5mC, providing evidence that TET1 in vivo oxidizes 5mC to 5hmC (Guo, Su, Zhong, Ming, & Song, 2011b; Kaas et al., 2013). The overexpression of Tet1 in the DG led to a significant decrease in methylation at promoter IX of *Bdnf* (brain-derived neurotrophic factor IX (*Bdnf IX*)) and the brain specific promoter of *Fgf1* (*Fgf1B*), and a concomitant increase in the expression of these two genes, supporting the role of Tet1 in the demethylation pathway, and subsequent gene activation (Guo et al., 2011b). Tet1 overexpression in area CA1 or DG of the hippocampus leads to the increased expression of various activity-dependent genes (*Fos*, *Arc*, *Egr1*, *Homer1*, and *Nf4a2*), as well as genes downstream of the Tet-mediated oxidation (*Tdg*, *Apobec1*, *Smug1*, and *Mbd4*) (Kaas et al., 2013). In the DG, the increased expression of these genes is dependent upon the catalytic domain TET1, as evidence by the fact that only the expression of human TET1 catalytic domain, but not expression of the catalytically inactive version of TET1; however, in the CA1 region, either the catalytic active or inactive TET1 leads to increase in expression of majority of these genes. This implies that TET1 acts via region-dependent mechanisms (Guo et al., 2011b; Kaas et al., 2013). Furthermore, Tet1 is required for neuronal activity–induced active DNA demethylation and gene expression since short hairpin–mediated knockdown of endogenous Tet1 in the DG abolished electroconvulsive stimulation (ECS)–induced demethylation of *Bdnf IX* and *Fgf1B* promoters. These in vivo findings are in agreement with in vitro work showing that Tet1 knockdown in hippocampal neurons leads to the hypermethylation of promoter IV of *Bdnf* and subsequent decreased expression from this promoter (Yu et al., 2015). Given that demethylation at these promoters is similarly abolished after ECS with knockdown of Apobec1, this suggests that TET1 and APOBEC1 work together through oxidative deamination to achieve active demethylation in the adult mouse brain (Guo et al., 2011b).

Loss of Tet1 also causes electrophysiological deficits in the hippocampus. Tet1 KO mice have normal basal synaptic transmission and intrinsic neuronal properties, as measured by paired-pulse facilitation and presynaptic excitability, respectively. However, long-term potentiation, assessed in the Schaffer collateral-CA1 pathway, is attenuated, and long-term depression (LTD) is amplified. These in vivo electrophysiological findings confirm what is found in vitro. Overexpression of the catalytically active form of TET1 prevents tetrodotoxin (TTX)-induced scaling-up, and knockdown of Tet1 leads to synaptic scaling-down that is unaltered by bicuculline treatment (Yu et al., 2015). Further analysis in vivo has demonstrated that alterations in the metabotropic glutamate receptor–dependent form of LTD is not affected, suggesting a deficit in *N*-methyl-D-aspartate receptor (NMDAR)–dependent LTD. Neuronal activity–regulated genes, including *c-Fos*, *Egr2*, *Egr4*, *Arc*, and *Npas4*, are affected in Tet1 KO mice. Analysis of the *Npas4* promoter-exon 1 region confirmed a decrease of 5hmC and an increase in 5mC, which could explain the downregulation of this group of genes. After memory extinction in Tet1 KO mice (but not after fear memory acquisition), the *Npas4* and *c-Fos* genes exhibit a decrease in 5hmC and an increase in 5mC, concomitant with a decrease in mRNA and protein expression levels in both brain regions assessed, the cortex and hippocampus. Since Tet1/Tet2/Tet3 expression does not increase during either fear memory extinction or acquisition, the activity of these proteins change, rather than absolute levels (Rudenko et al., 2013).

This body of work on Tet1 function in the brain suggests that Tet1, although expressed at much lower levels in the mammalian brain than the other Tet family members, plays an important role in maintaining 5hmC levels, and subsequent gene expression levels, at basal and activity-induced conditions.

TET2

Despite its high level of expression, Tet2 is presently the least well-studied Tet family member in the brain, but the limited studies conducted thus far have demonstrated the importance of Tet2 in brain function. There are no reported brain abnormalities or dysfunction in Tet2 KO mice (Ko et al., 2011; Li et al., 2011). However, when Tet2 is knocked down in hippocampal neurons in vitro, there is an increase in miniature excitatory postsynaptic current (mEPSC) amplitudes compared to controls (Yu et al., 2015). This implies that neuronal function may be impaired in the absence of Tet2.

Tet2 is also thought to play a role in the demethylation of developmentally dependent genomic loci. With the use of Tet2 KO mice, it was found that this member of the Tet family is responsible for the oxidation of large fraction (19.7%) of CpG genomic regions that gain hydroxymethylation status over development. In contrast, CpG regions with higher 5hmC in the adult than fetal stage are largely unaffected in Tet2 KO mice. Across development and aging (6 week, 10 week, and 22 mo) in Tet2 KO mice, there are greater than four-fold more hypermethylated CpG regions (14,000 CpG regions in total) than hypomethylated regions, suggesting that Tet2 plays a role in the demethylation over development and aging (Lister et al., 2013). Tet2 may also play a role in neurogenesis since the knockdown of Tet2 and Tet3 via electroporation of shRNAs into the cortex lead to defects in the progression of differentiation from the subventricular zone (Hahn et al., 2013). These Tet2 findings suggest that Tet2 plays an important role in regulating developmentally dependent, differentially hydroxymethylated regions.

TET3

Various studies on Tet3 function in the brain have confirmed that this most highly expressed Tet family member in the brain is essential for regulating neuronal activity. When mice undergo extinction training, there is a significant increase in Tet3 mRNA in the cortex. Tet3 knockdown via lentiviral plasmids in the infralimbic prefrontal cortex results in normal fear memory acquisition, but impaired fear memory extinction. Furthermore, inhibiting NMDAR activity blocks the increase in Tet3 expression associated with fear memory extinction, suggesting that the rise in Tet3 occurs via an NMDAR-mediated pathway. Fear extinction causes in genome-wide changes in 5hmC at locations that contain CpA or CpT dinucleotide repeats, but not at CpGs. Additionally, there is a reduction in 5hmC at intronic and intergenic sites and an increase in 5hmC enrichment at distal promoters, 5′ untranslated region (UTR), 3′ UTR, exonic sequences, and DHS regions. Gene ontology analysis revealed that 16% genes enriched for 5hmC after extinction learning are involved in synaptic signaling. When one of these genes, gephyrin (*Gphn*), was evaluated, it was found that there was enrichment for 5hmC, co-occurring with decrease in 5mC, within one intron. Moreover, in response to extinction, there was an increase Tet3 occupancy at *Gphn*, as well as an increase in specificity protein 1 (*Sp1*), a transcription factor that activates gene expression by preventing the active loci from becoming methylated. Additionally, the observed reduction in transient H3K9me3 and increase in H3K27ac, p300, H3K4me1, and di-methyl arginine of histone H3 (H3R2me2), which support a euchromatic state, support the role of Tet3 in extinction-induced gene expression changes. All of these changes at the *Gphn* appear to be specifically regulated by Tet3 function, since the changes are blocked with the use of a Tet3 shRNA (Li et al., 2014).

Tet3 expression levels correlate with neuronal activity in vitro as well; an increase in synaptic transmission correlates with an increase in Tet3, but not Tet1 or Tet2. When Tet3 is knocked down from hippocampal neurons in culture, mEPSC amplitudes are significantly larger than controls, and the reciprocal effect occurs when Tet3 is overexpressed. Notably, knockdown of either Tet1 or Tet2 also increases mEPSC amplitudes, but not as drastically as Tet3 knockdown. Tet3 is also essential for the maintenance of homeostatic synaptic plasticity since knockdown of Tet3 leads to synaptic scaling-up that is unaltered by TTX or retinoic acid treatment; knockdown of Tet3 leads to synaptic scaling-down that occludes further alterations with bicuculline treatment; and Tet3 overexpression prevents TTX-induced synaptic scaling-up or bicuculline-induced scaling-down. Given that a similar effect on mEPSC amplitudes and synaptic scaling occurs when poly(ADP-ribose) polymerase or apurinic/apyrimidinic endonuclease, the two major components of the BER pathway, is inhibited this suggests that excitatory synaptic transmission is regulated by the oxidation of DNA via TET, followed by BER (Yu et al., 2015).

The molecular mechanism through which Tet3 elicits these effects is likely through the regulation of surface glutamate receptor 1 (GluR1). Knockdown of Tet3 leads to an increase in surface GluR1 receptors that is resistant to a further increase or decrease in surface GluR1 receptors. When gene expression changes were assessed in Tet3 knockdown neurons, gene ontology term enrichment revealed expression changes of genes involved in the synapse and synaptic transmission. Genes with differential expression due to TTX or bicuculline treatment in control neurons lost responsiveness in Tet3 knockdown neurons. In Tet3 knockdown neurons, promoter IV of the *Bdnf* is hypermethylated, and there is a decrease in expression from this promoter. The bicuculline-induced hypomethylation, as well as the TTX-induced hypermethylation, of *Bdnf* promoter IV is occluded in Tet3 knockdown neurons. Chromatin immunoprecipitation (ChIP)-PCR revealed that TET3 binds to the *Bdnf* promoter IV (Yu et al., 2015). These findings together suggest that neuronal activity can regulate TET3 function, which in turn controls the expression of target genes via altering methylation levels.

The importance of Tet3 in neural function is conserved across vertebrates, as knockdown of Tet3 in *Xenopus* by morpholino antisense oligonucleotide leads to marked developmental abnormalities, including malformation of the eye, small head, and early death. At the molecular level, Tet3 depletion causes a reduction in expression of master eye developmental genes (*pax6*, *rx*, and *six3*), primary neuronal markers (*ngn2* and *tubb2b*), neural crest markers (*sox9* and *snail*), and major sonic hedgehog (shh) signaling components (*shh* and *ptc-1*). Additionally, TET3-ChIP assays confirm the binding of Tet3 to the promoters of *pax6*, *rx*, *six3*, *ptc-1*, *ptc-2*, *sox9*, and *ngn2*. Furthermore, at the promoters of some of these target genes, there was found to be a developmental increase in 5hmC and decrease in 5mC from stage 10 to 19 in *Xenopus* development, which is perturbed when Tet3 is knocked down. These findings suggest that TET3 acts as an upstream activator of key neural developmental genes (Xu et al., 2012). Furthermore, these studies suggest that TET3 plays an important role in brain function that is conserved across animals.

CONCLUSION

Despite its recent rediscovery in 2009, major advances have been made regarding our understanding of the DNA modification 5hmC. 5hmC is now known to be an intermediate of active DNA demethylation in neurons. However, the biological role of 5hmC and its regulation in the brain are still debated. It is known that even within defined brain regions, such as the cerebellum, there are a multitude of cell types that have vastly different functions, gene expression patterns, and 5mC and 5hmC levels. Although studies have shown variable 5hmC distribution in different brain regions throughout development, it is still not understood what happens to 5hmC in specific cell types as an animal ages. Investigating 5hmC distribution in a heterogeneous brain region is likely to mask important cell type–specific features and reveal false patterns due to the multitude of cell types in the sample. Looking at 5hmC and Tet enzymes in specific cell types is crucial for our understanding of how DNA demethylation and 5hmC affect neural functions.

In the search for the biological role of 5hmC, it is also imperative to know what proteins bind to the modified base, and the cellular consequences of the interactions. Although a variety of putative 5hmC binding partners have been identified, the potential interactions have not been confirmed in vivo, and the functions of 5hmC–protein interactions have not been established. Studies addressing the binding partners of 5hmC will greatly advance our understanding of 5hmC not only as a demethylation intermediate but also as an essential epigenetic mark.

REFERENCES

Baubec, T., Ivanek, R., Lienert, F., & Schübeler, D. (2013). Methylation-dependent and -independent genomic targeting principles of the MBD protein family. *Cell, 153*(2), 480–492. http://dx.doi.org/10.1016/j.cell.2013.03.011.

Blaschke, K., Ebata, K. T., Karimi, M. M., Zepeda-Martínez, J. A., Goyal, P., Mahapatra, S., et al. (2013). Vitamin C induces Tet-dependent DNA demethylation and a blastocyst-like state in ES cells. *Nature, 500*(7461), 222–226. http://dx.doi.org/10.1038/nature12362.

Chen, H., Dzitoyeva, S., & Manev, H. (2012). Effect of aging on 5-hydroxymethylcytosine in the mouse hippocampus. *Restorative Neurology and Neuroscience, 30*(3), 237–245. http://dx.doi.org/10.3233/RNN-2012-110223.

Chouliaras, L., van den Hove, D. L. A., Kenis, G., Keitel, S., Hof, P. R., van Os, J., et al. (2012). Age-related increase in levels of 5-hydroxymethylcytosine in mouse hippocampus is prevented by caloric restriction. *Current Alzheimer Research, 9*(5), 536–544.

Chouliaras, L., Mastroeni, D., Delvaux, E., Grover, A., Kenis, G., Hof, P. R., et al. (2013). Consistent decrease in global DNA methylation and hydroxymethylation in the hippocampus of Alzheimer's disease patients. *Neurobiology of Aging, 34*, 2091–2099. http://dx.doi.org/10.1016/j.neurobiolaging.2013.02.021.

Colquitt, B. M., Allen, W. E., Barnea, G., & Lomvardas, S. (2013). Alteration of genic 5-hydroxymethylcytosine patterning in olfactory neurons correlates with changes in gene expression and cell identity. *Proceedings of the National Academy of Sciences of the United States of America, 110*(36), 14682–14687. http://dx.doi.org/10.1073/pnas.1302759110.

Condliffe, D., Wong, A., Troakes, C., Proitsi, P., Patel, Y., Choiliaras, L., et al. (2014). Cross-region reduction in 5-hydroxymethylcytosine in Alzheimer's disease brain. *Neurobiology of Aging, 35*, 1850–1854. http://dx.doi.org/10.1016/j.neurobiolaging.2014.02.002.

Coppieters, N., Dieriks, B. V., Lill, C., Faull, R. L. M., Curtis, M. A., & Dragunow, M. (2014). Global changes in DNA methylation and hydroxymethylation in Alzheimer's disease human brain. *Neurobiology of Aging, 35*, 1334–1344. http://dx.doi.org/10.1016/j.neurobiolaging.2013.11.031.

Dawlaty, M. M., Breiling, A., Le, T., Raddatz, G., Barrasa, M. I., Cheng, A. W., et al. (2013). Combined deficiency of Tet1 and Tet2 causes epigenetic abnormalities but is compatible with postnatal development. *Developmental Cell, 24*(3), 310–323. http://dx.doi.org/10.1016/j.devcel.2012.12.015.

Dawlaty, M. M., Ganz, K., Powell, B. E., Hu, Y.-C., Markoulaki, S., Cheng, A. W., et al. (2011). Tet1 is dispensable for maintaining pluripotency and its loss is compatible with embryonic and postnatal development. *Cell Stem Cell, 9*(2), 166–175. http://dx.doi.org/10.1016/j.stem.2011.07.010.

Frauer, C., Hoffmann, T., Bultmann, S., Casa, V., Cardoso, M. C., Antes, I., et al. (2011). Recognition of 5-hydroxymethylcytosine by the Uhrf1 SRA domain. *PLoS One, 6*(6), e21306. http://dx.doi.org/10.1371/journal.pone.0021306.

Globisch, D., Münzel, M., Müller, M., Michalakis, S., Wagner, M., Koch, S., et al. (2010). Tissue distribution of 5-hydroxymethylcytosine and search for active demethylation intermediates. *PLoS One, 5*(12), e15367. http://dx.doi.org/10.1371/journal.pone.0015367.

Guo, J. U., Ma, D. K., Mo, H., Ball, M. P., Jang, M.-H., Bonaguidi, M. A., et al. (2011a). Neuronal activity modifies the DNA methylation landscape in the adult brain. *Nature Neuroscience, 14*(10), 1345–1351. http://dx.doi.org/10.1038/nn.2900.

Guo, J. U., Su, Y., Zhong, C., Ming, G.-L., & Song, H. (2011b). Hydroxylation of 5-methylcytosine by TET1 promotes active DNA demethylation in the adult brain. *Cell, 145*(3), 423–434. http://dx.doi.org/10.1016/j.cell.2011.03.022.

Hahn, M. A., Qiu, R., Wu, X., Li, A. X., Zhang, H., Wang, J., et al. (2013). Dynamics of 5-hydroxymethylcytosine and chromatin marks in mammalian neurogenesis. *Cell Reports, 3*(2), 291–300. http://dx.doi.org/10.1016/j.celrep.2013.01.011.

He, Y.-F., Li, B.-Z., Li, Z., Liu, P., Wang, Y., Tang, Q., et al. (2011). Tet-mediated formation of 5-carboxylcytosine and its excision by TDG in mammalian DNA. *Science (New York, NY)*, *333*(6047), 1303–1307. http://dx.doi.org/10.1126/science.1210944.

Ito, S., Shen, L., Dai, Q., Wu, S. C., Collins, L. B., Swenberg, J. A., et al. (2011). Tet proteins can convert 5-methylcytosine to 5-formylcytosine and 5-carboxylcytosine. *Science (New York, NY)*, *333*(6047), 1300–1303. http://dx.doi.org/10.1126/science.1210597.

Iyer, L. M., Tahiliani, M., Rao, A., & Aravind, L. (2009). Prediction of novel families of enzymes involved in oxidative and other complex modifications of bases in nucleic acids. *Cell Cycle (Georgetown, Tex.)*, *8*(11), 1698–1710.

Jaenisch, R., & Bird, A. (2003). Epigenetic regulation of gene expression: how the genome integrates intrinsic and environmental signals. *Nature Genetics*, *33*(Suppl.), 245–254. http://dx.doi.org/10.1038/ng1089.

Jin, S.-G., Wu, X., Li, A. X., & Pfeifer, G. P. (2011). Genomic mapping of 5-hydroxymethylcytosine in the human brain. *Nucleic Acids Research*, *39*(12), 5015–5024. http://dx.doi.org/10.1093/nar/gkr120.

Kaas, G. A., Zhong, C., Eason, D. E., Ross, D. L., Vachhani, R. V., Ming, G.-L., et al. (2013). TET1 controls CNS 5-methylcytosine hydroxylation, active DNA demethylation, gene transcription, and memory formation. *Neuron*, *79*(6), 1086–1093. http://dx.doi.org/10.1016/j.neuron.2013.08.032.

Khare, T., Pai, S., Koncevicius, K., Pal, M., Kriukiene, E., Liutkeviciute, Z., et al. (2012). 5-hmC in the brain is abundant in synaptic genes and shows differences at the exon-intron boundary. *Nature Structural & Molecular Biology*, *19*(10), 1037–1043. http://dx.doi.org/10.1038/nsmb.2372.

Khrapunov, S., Warren, C., Cheng, H., Berko, E. R., Greally, J. M., & Brenowitz, M. (2014). Unusual characteristics of the DNA binding domain of epigenetic regulatory protein MeCP2 determine its binding specificity. *Biochemistry*, *53*(21), 3379–3391. http://dx.doi.org/10.1021/bi500424z.

Ko, M., Bandukwala, H. S., An, J., Lamperti, E. D., Thompson, E. C., Hastie, R., et al. (2011). Ten-Eleven-Translocation 2 (TET2) negatively regulates homeostasis and differentiation of hematopoietic stem cells in mice. *Proceedings of the National Academy of Sciences of the United States of America*, *108*(35), 14566–14571. http://dx.doi.org/10.1073/pnas.1112317108.

Kohli, R. M., & Zhang, Y. (2013). TET enzymes, TDG and the dynamics of DNA demethylation. *Nature*, *502*(7472), 472–479. http://dx.doi.org/10.1038/nature12750.

Kriaucionis, S., & Heintz, N. (2009). The nuclear DNA base 5-hydroxymethylcytosine is present in Purkinje neurons and the brain. *Science (New York, NY)*, *324*(5929), 929–930. http://dx.doi.org/10.1126/science.1169786.

Li, Z., Cai, X., Cai, C.-L., Wang, J., Zhang, W., Petersen, B. E., et al. (2011). Deletion of Tet2 in mice leads to dysregulated hematopoietic stem cells and subsequent development of myeloid malignancies. *Blood*, *118*(17), 4509–4518. http://dx.doi.org/10.1182/blood-2010-12-325241.

Li, X., Wei, W., Zhao, Q.-Y., Widagdo, J., Baker-Andresen, D., Flavell, C. R., et al. (2014). Neocortical Tet3-mediated accumulation of 5-hydroxymethylcytosine promotes rapid behavioral adaptation. *Proceedings of the National Academy of Sciences of the United States of America*, *111*(19), 7120–7125. http://dx.doi.org/10.1073/pnas.1318906111.

Lister, R., Mukamel, E. A., Nery, J. R., Urich, M., Puddifoot, C. A., Johnson, N. D., et al. (2013). Global epigenomic reconfiguration during mammalian brain development. *Science (New York, NY)*. http://dx.doi.org/10.1126/science.1237905.

Maunakea, A. K., Chepelev, I., Cui, K., & Zhao, K. (2013). Intragenic DNA methylation modulates alternative splicing by recruiting MeCP2 to promote exon recognition. *Cell Research*, *23*(11), 1256–1269. http://dx.doi.org/10.1038/cr.2013.110.

Mellén, M., Ayata, P., Dewell, S., Kriaucionis, S., & Heintz, N. (2012). MeCP2 binds to 5hmC enriched within active genes and accessible chromatin in the nervous system. *Cell*, *151*(7), 1417–1430. http://dx.doi.org/10.1016/j.cell.2012.11.022.

Münzel, M., Globisch, D., Brückl, T., Wagner, M., Welzmiller, V., Michalakis, S., et al. (2010). Quantification of the sixth DNA base hydroxymethylcytosine in the brain. *Angewandte Chemie (International ed. in English)*, *49*(31), 5375–5377. http://dx.doi.org/10.1002/anie.201002033.

Penn, N. W., Suwalski, R., O'Riley, C., Bojanowski, K., & Yura, R. (1972). The presence of 5-hydroxymethylcytosine in animal deoxyribonucleic acid. *The Biochemical Journal*, *126*(4), 781–790.

Rudenko, A., Dawlaty, M. M., Seo, J., Cheng, A. W., Meng, J., Le, T., et al. (2013). Tet1 is critical for neuronal activity-regulated gene expression and memory extinction. *Neuron*, *79*(6), 1109–1122. http://dx.doi.org/10.1016/j.neuron.2013.08.003.

Song, C.-X., Szulwach, K. E., Fu, Y., Dai, Q., Yi, C., Li, X., et al. (2011). Selective chemical labeling reveals the genome-wide distribution of 5-hydroxymethylcytosine. *Nature Biotechnology*, *29*(1), 68–72. http://dx.doi.org/10.1038/nbt.1732.

Spruijt, C. G., Gnerlich, F., Smits, A. H., Pfaffeneder, T., Jansen, P. W. T. C., Bauer, C., et al. (2013). Dynamic readers for 5-(hydroxy)methylcytosine and its oxidized derivatives. *Cell*, *152*(5), 1146–1159. http://dx.doi.org/10.1016/j.cell.2013.02.004.

Sun, W., Zang, L., Shu, Q., & Li, X. (2014). From development to diseases: the role of 5hmC in brain. *Genomics*, *104*(5), 347–351. http://dx.doi.org/10.1016/j.ygeno.2014.08.021.

Szulwach, K. E., Li, X., Li, Y., Song, C.-X., Wu, H., Dai, Q., et al. (2011). 5-hmC-mediated epigenetic dynamics during postnatal neurodevelopment and aging. *Nature Neuroscience*, *14*(12), 1607–1616. http://dx.doi.org/10.1038/nn.2959.

Szwagierczak, A., Bultmann, S., Schmidt, C. S., Spada, F., & Leonhardt, H. (2010). Sensitive enzymatic quantification of 5-hydroxymethylcytosine in genomic DNA. *Nucleic Acids Research*, *38*(19), e181. http://dx.doi.org/10.1093/nar/gkq684.

Tahiliani, M., Koh, K. P., Shen, Y., Pastor, W. A., Bandukwala, H., Brudno, Y., et al. (2009). Conversion of 5-methylcytosine to 5-hydroxymethylcytosine in mammalian DNA by MLL partner TET1. *Science (New York, NY)*, *324*(5929), 930–935. http://dx.doi.org/10.1126/science.1170116.

UniProt Consortium. (2015). UniProt: a hub for protein information. *Nucleic Acids Research*, *43*(Database issue), D204–D212. http://dx.doi.org/10.1093/nar/gku989.

Valinluck, V., Tsai, H.-H., Rogstad, D. K., Burdzy, A., Bird, A., & Sowers, L. C. (2004). Oxidative damage to methyl-CpG sequences inhibits the binding of the methyl-CpG binding domain (MBD) of methyl-CpG binding protein 2 (MeCP2). *Nucleic Acids Research*, *32*(14), 4100–4108. http://dx.doi.org/10.1093/nar/gkh739.

Wang, F., Yang, Y., Lin, X., Wang, J.-Q., Wu, Y.-S., Xie, W., et al. (2013). Genome-wide loss of 5-hmC is a novel epigenetic feature of Huntington's disease. *Human Molecular Genetics*, *22*(18), 3641–3653. http://dx.doi.org/10.1093/hmg/ddt214.

Wang, Y., & Zhang, Y. (2014). Regulation of TET protein stability by calpains. *Cell Reports*, *6*(2), 278–284. http://dx.doi.org/10.1016/j.celrep.2013.12.031.

Wen, L., Li, X., Yan, L., Tan, Y., Li, R., Zhao, Y., et al. (2014). Whole-genome analysis of 5-hydroxymethylcytosine and 5-methylcytosine at base resolution in the human brain. *Genome Biology*, *15*(3), R49. http://dx.doi.org/10.1186/gb-2014-15-3-r49.

Wheldon, L. M., Abakir, A., Ferjentsik, Z., Dudnakova, T., Strohbuecker, S., Christie, D., et al. (2014). Transient accumulation of 5-carboxylcytosine indicates involvement of active demethylation in lineage specification of neural stem cells. *Cell Reports*, *7*(5), 1353–1361. http://dx.doi.org/10.1016/j.celrep.2014.05.003.

Wyatt, G. R., & Cohen, S. S. (1953). The bases of the nucleic acids of some bacterial and animal viruses: the occurrence of 5-hydroxymethylcytosine. *The Biochemical Journal*, *55*(5), 774–782.

Xu, Y., Xu, C., Kato, A., Tempel, W., Abreu, J. G., Bian, C., et al. (2012). Tet3 CXXC domain and dioxygenase activity cooperatively regulate key genes for Xenopus eye and neural development. *Cell*, *151*(6), 1200–1213. http://dx.doi.org/10.1016/j.cell.2012.11.014.

Yin, R., Mao, S.-Q., Zhao, B., Chong, Z., Yang, Y., Zhao, C., et al. (2013). Ascorbic acid enhances Tet-mediated 5-methylcytosine oxidation and promotes DNA demethylation in mammals. *Journal of the American Chemical Society*, *135*(28), 10396–10403. http://dx.doi.org/10.1021/ja4028346.

Yu, M., Hon, G. C., Szulwach, K. E., Song, C.-X., Zhang, L., Kim, A., et al. (2012). Base-resolution analysis of 5-hydroxymethylcytosine in the mammalian genome. *Cell*, *149*(6), 1368–1380. http://dx.doi.org/10.1016/j.cell.2012.04.027.

Yu, H., Su, Y., Shin, J., Zhong, C., Guo, J. U., Weng, Y.-L., et al. (2015). Tet3 regulates synaptic transmission and homeostatic plasticity via DNA oxidation and repair. *Nature Neuroscience*, *18*(6), 836–843. http://dx.doi.org/10.1038/nn.4008.

Zhang, R.-R., Cui, Q.-Y., Murai, K., Lim, Y. C., Smith, Z. D., Jin, S., et al. (2013). Tet1 regulates adult hippocampal neurogenesis and cognition. *Cell Stem Cell*, *13*(2), 237–245. http://dx.doi.org/10.1016/j.stem.2013.05.006.

CHAPTER 5

Beyond mCG: DNA Methylation in Noncanonical Sequence Context

E.A. Mukamel[1], R. Lister[2,3]
[1]University of California San Diego, La Jolla, CA, United States; [2]The University of Western Australia, Perth, WA, Australia; [3]The Harry Perkins Institute of Medical Research, Perth, WA, Australia

INTRODUCTION: BEYOND CG METHYLATION

The genetic code based on the sequence of DNA nucleotides A, C, G, and T is thought to be a universal and invariant feature of life on Earth. Advances in genome sequencing have enabled comprehensive approaches to epigenome profiling that increasingly point to a diverse set of extensions to the genomic code in different cell types, even within the same species. Symbolic diversity is a familiar feature of human languages: French rarely uses the letters *k* or *w*, but it has access to a range of diacritical marks such as *é* that are not used in English. Just as we can easily recognize *walked* as English and *marché* as French based on the symbols in each word, recent epigenomic profiling efforts allow a new appreciation of how biological cells and tissues use distinct alphabets of epigenomic marks to regulate their specialized cellular functions. The new findings increasingly show that the mammalian epigenome is a symbolic system that encodes, stores, and transmits information through development and, potentially, across generations. Epigenomic marks, including DNA methylation and covalent modification of histone proteins, enhance the coding capacity of the genome by expanding the number of symbols available for representing gene regulatory information. To understand genomic information processing, we must make sense of the variety of symbolic elements used by particular cell types in each species.

Methylation of cytosine in genomic DNA is an essential epigenetic modification that primarily represses transcription and regulates other genomic processes across most, although not all, plant and animal species. The widespread presence and functional role of methylcytosine at CG dinucleotides has long been recognized (Suzuki & Bird, 2008); however, methylation at CA, CT, and CC positions (collectively called non–CG methylation, or mCH) in mammalian cells has also been established (Ramsahoye et al., 2000). By combining modern whole-genome shotgun DNA sequencing with sodium bisulfite conversion, in a technique called MethylC-seq (Lister & Ecker, 2009), the methylation status of more than 90% of genomic cytosines can now be experimentally determined at single-base resolution. Although MethylC-seq detects both methyl- and hydroxymethyl-cytosine (mC and hmC), techniques

DNA Modifications in the Brain
ISBN 978-0-12-801596-4
http://dx.doi.org/10.1016/B978-0-12-801596-4.00005-8

81

such as Tet-assisted bisulfite sequencing (TAB-seq) profiling (Yu et al., 2012) enable the two modifications to be distinguished at base resolution throughout the genome.

This advance in methylome profiling technology first showed that although IMR90 human fetal lung fibroblast cells contain <0.02% of their methylcytosine in the non–CG context, human embryonic stem (ES) cells harbor nearly a quarter of their methylcytosines at non–CG positions (Lister et al., 2009). Subsequent surveys of a range of cells initially seemed to confirm this pattern, showing abundant non–CG methylation in pluripotent cells (Laurent et al., 2010; Lister et al., 2011), but little or no non–CG methylation across differentiated cell types including primary tissue samples and differentiated cells derived from pluripotent cells (Xie et al., 2013; Ziller et al., 2011). It was surprising, then, that MethylC-seq profiling of brain tissue from mouse (Xie et al., 2012) and human (Lister et al., 2013; Varley et al., 2013; Zeng et al., 2012) revealed a substantial amount of non–CG methylation. By purifying nuclei of neurons expressing the marker NeuN, cell type–specific profiling showed that non–CG methylation accounts for roughly half of all methylcytosine in adult frontal cortex neurons (Lister et al., 2013). This represents the most abundant level of non–CG methylation of any cell type yet observed. TAB-seq profiling in mouse frontal cortex and human ES cells showed that almost all of the non–CG methylation is in the form of mC and not hmC (Lister et al., 2013; Yu et al., 2012).

To appreciate the potential significance of non–CG methylation, it is important to consider the density of CG and non–CG positions in the human genome (Fig. 5.1). CG

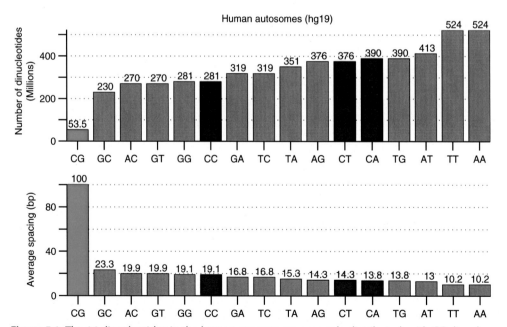

Figure 5.1 The 16 dinucleotides in the human genome are unevenly distributed, with CG dinucleotides (green) greatly depleted compared with non–CG positions (blue). As a result, the average spacing between CG positions is ~100 bp, whereas non–CG positions occur every ~2.1 bp.

dinucleotides occur at around 1 in 100 positions in the genome, far less than the 1 in 16 positions expected in a random sequence. CG sequences have been lost during evolution due to the higher rate of mutation of methylcytosine (Saxonov, Berg, & Brutlag, 2006). Around 11% of CG positions are concentrated in CG islands, a small genomic compartment associated with gene promoters and covering ~1.4% of the genome. By contrast, non–CG positions occur at approximately 1 of every 2.1 bp. Thus, even a low rate of methylation at non–CG sites may have a substantial impact by virtue of the 50-fold increased density of these sites relative to CG positions.

VARIABLE NON–CG METHYLATION ACROSS CELL TYPES

Cellular differentiation requires extensive changes in the use of genomic information without changes to the underlying DNA sequence. Epigenome remodeling accompanies the regulation of gene transcription that defines and maintains the identity of specialized cell types. Although cells throughout the body share the same genetic sequence, apart from limited somatic mutations (Baillie et al., 2011; McConnell et al., 2013; Upton et al., 2015), it is the genome-wide pattern of DNA methylation and chromatin modifications that provide a cell type–specific fingerprint of each cell type's use of the genetic information. Whereas the cell type–specific transcriptome is a snapshot of a cell's current biological state, epigenomic information may reflect the past, current, and potential future dynamical regulation (Hon et al., 2013). For example, key neuronal transcription factors such as Npas4, which plays distinct cell type–specific roles in excitatory and inhibitory cortical neurons (Spiegel et al., 2014), are transcribed in response to neuronal depolarization or synaptic input. Cell type–specific epigenomic modifications likely regulate these activity-dependent responses. It is thus critical to understand differences in the epigenomic modifications present in different cell types to elucidate the role they play in cellular specialization.

The presence of dramatically elevated non–CG methylation levels in neurons suggests that this epigenetic mark could be present in one or more of the myriad other classes of human cells. As part of the NIH Roadmap Epigenomics project, MethylC-seq was used to profile the DNA methylomes of 18 human tissues from 36 samples (Schultz et al., 2015). Neurons were found to contain the highest rate of non–CG methylation, far more than any other cell type, with nearly 10% of all non–CG sites methylated. Glia and ES cells were also relatively highly methylated in the non–CG context, although they possess levels far lower than those observed in mature neurons (Lister et al., 2013; Schultz et al., 2015). Notably, the abundance of non–CG methylation in mammalian neurons changes dramatically during brain development (Lister et al., 2013). Negligible non–CG methylation is detected in the fetal mammalian brain, however a rapid accumulation of non–CG methylation occurs during postnatal brain development, specifically between 1 and 4 weeks after birth in mice, and in the first 2 years after birth in humans.

Although Schultz et al. (2015) also demonstrated that non–CG methylation is detectable throughout the genome across most of the tissues profiled, its level is much lower than in neural tissue and pluripotent cells, with the next highest enrichment observed in heart, muscle, and bladder (0.2–0.37%). The pattern of non–CG methylation across these tissues parallels that which occurs in neurons and glia: relatively lower methylation in gene bodies of actively transcribed genes and higher methylation in the bodies of repressed genes. Based on the profile of non–CG methylation, clusters of genes with muscle- or heart-specific non–CG methylation patterns could also be identified (Schultz et al., 2015). In particular, these clusters were enriched for genes with tissue-specific functional annotations.

Embryonic stem cells also harbor a substantial amount of non–CG methylation (Lister et al., 2009). This signature is a hallmark of early pluripotency that is shared by induced pluripotent stem cells (Lister et al., 2011; Ziller et al., 2011). Non–CG methylation is rapidly lost upon differentiation of ES cells to a range of lineages, including neural progenitor cells (Xie et al., 2013). However, the functional profile of non–CG methylation in pluripotent cells differs markedly from adult tissues. Whereas the density of non–CG methylation in gene bodies correlates with transcriptional repression in differentiated tissues and cell types, exactly the opposite pattern prevails in pluripotent cells (Lister et al., 2013, 2009; Ziller et al., 2011). Hypermethylation of gene bodies of actively transcribed genes persists in cell lineages derived through in vitro differentiation of ES cells, including mesendoderm, trophoblast, neural progenitor cells, and mesenchymal stem cells (Xie et al., 2013). Although these cell lineages have lower levels of non–CG methylation compared with ES cells, the genomic distribution of this mark is similar across each of these lineages, but highly distinct from adult tissues (Schultz et al., 2015).

The distinct regulation of non–CG methylation in differentiated cell types compared with pluripotent cells is also reflected in their different local sequence contexts. Non–CG methylation occurs at CA and CT positions, with many different flanking sequences, but it is all but undetectable at CC positions. Among all CA and CT positions, methylcytosine is most enriched at CAC sites in neurons and in all differentiated cell types and tissues. In contrast, CAG is highly enriched among methylated non–CG positions in pluripotent cells and lineages derived from them (Lister et al., 2013; Schultz et al., 2015; Varley et al., 2013). This distinct sequence context suggests, at least within these two broad classes of cells, that the methyltransferases responsible for depositing non–CG methylation are either distinct or modulated by different cofactors or posttranslational modifications. Indeed, the de novo methyltransferase DNMT3A is necessary for CA and CT methylation in mouse brain (Gabel et al., 2015; Guo et al., 2014), and, together with DNMT3B, it is responsible for establishing non–CG methylation patterns during early embryogenesis (Okano, Bell, Haber, & Li, 1999) and non–CG methylation in pluripotent cell types (Ziller et al., 2011). Pluripotent cells, but not brain tissue, express the cofactor DNMT3L, which lacks methyltransferase activity but is able to mediate the recruitment of the DNMT3A/B-DNMT3L complex to nucleosomes harboring unmethylated histone H3 at lysine 4 (H3K4) (Ooi et al., 2007). DNMT3L is required for the establishment of imprinting during early development (Bourc'his, Xu, Lin, Bollman, & Bestor, 2001). Furthermore, DNMT3B, which is expressed at very low levels in the brain compared to DNMT3A, possesses a PWWP domain that mediates its recruitment to H3K36me3, a chromatin modification that is abundant in the body of actively transcribed genes (Baubec et al., 2015). Thus, these differences in the methylation machinery may explain the distinct sequence contexts of non–CG methylation in differentiated compared with pluripotent cells. However, it remains unclear why non–CG methylation shows the opposite association with transcription in these two cell classes.

In addition to non–CG methylation being an effective marker for genes that exhibit particular states of transcriptional activity in pluripotent cells and the brain, large regions of non–CG methylation enrichment or depletion have been identified as effective markers of various functional states in different cell types. In the brain, large genomic regions that are almost completely devoid of non–CG methylation, referred to as mCH deserts, exhibit highly inaccessible chromatin states that seem to resist de novo methylation, and are enriched for olfactory receptor gene and immunoglobulin gene clusters (Lister et al., 2013). In human induced pluripotent stem (iPS) cells, megabase-scale regions of the genome frequently fail to regain non–CG methylation during the reprogramming process and remain hypomethylated compared to ES cells in the non–CG context. These extensive differentially methylated regions (non–CG mega-DMRs) tend to occur at genomic regions which, in the differentiated cells from which the iPS cells were reprogrammed, exist in a partially methylated state in the CG context, referred to as partially methylated domains (PMDs) (Lister et al., 2009, 2011). PMDs are associated with late replicating genomic regions that localize to the nuclear lamina (Berman et al., 2012). They display low transcriptional activity and harbor repressive chromatin modifications such as H3K9me3 (Lister et al., 2009). Non–CG mega-DMRs in iPS cells frequently display high levels of H3K9me3, in contrast to the same regions in ES cells, and serve as highly effective epigenomic markers that allow discrimination of iPS cells from ES cells (Lister et al., 2011). Furthermore, as discussed in further detail below, female genes that escape X chromosome inactivation are marked by hypermethylation in the non–CG context, which allows effective identification of this unique regulatory state (Lister et al., 2013; Schultz et al., 2015). Thus, the pattern of non–CG methylation serves as a highly effective marker for cellular identity and genome regulatory states that can be assessed simply from a genomic DNA sample.

NON–CG METHYLATION AND BRAIN CELL DIVERSITY

Maintaining a diverse population of specialized cell types is important in most tissues, and it is especially critical for the function of brain circuits. Cognitive processes such as perception, memory, and motor control rely on the balanced interaction, via coordinated gene expression, protein synthesis, and synaptic signaling, of a myriad of excitatory and inhibitory neuron types. Distinct neuronal cell types express unique DNA-binding transcription factors, have different histone modification profiles, and also differ substantially in their landscape of CG and non–CG DNA methylation. Among these cell type–specific characteristics, the pattern of non–CG methylation is one of the most distinctive molecular-genetic fingerprints of cell type–specific iden-tity. In a comparison of three neuronal cell types (excitatory pyramidal cells, and inhibitory cells expressing parvalbumin or vasoactive intestinal peptide), almost half of all protein-coding genes were found to harbor neuron type–specific non–CG meth-ylation (Mo et al., 2015). The correlation of gene body mCH levels between neuronal cell types ($r = 0.83–0.86$) was lower than the corresponding correlation of transcrip-tional levels by RNA-seq ($r = 0.95–0.96$), again suggesting that mCH captures essen-tial aspects of neuronal epigenetic diversity. By contrast, both CG and non–CG methylation are precisely conserved in replicate experiments using tissue samples from different individuals, and non–CG methylation is conserved at homologous sequences between human and mouse brain neurons (Lister et al., 2013).

In cell types that harbor non–CG methylation, the mark appears across almost all genomic compartments, including exons, introns, and outside of gene bodies. Active regulatory elements located outside of gene promoters, as defined by a suite of epigen-etic and chromatin marks, are depleted of non–CG methylation (Lister et al., 2009; Mo et al., 2015). DMRs, which are classically defined based on statistically significant differ-ences in the level of CG methylation, are also marked by differential non–CG methyla-tion as well as cell type–specific active histone marks (Lister et al., 2013; Mo et al., 2015). These active regulatory regions are also marked by open chromatin as assayed by in vitro transposition of native chromatin by Tn5 transposase (ATAC-seq) (Buenrostro, Giresi, Zaba, Chang, & Greenleaf, 2013). By computationally identifying ATAC-seq footprints lying over transcription factor sequence motifs, putative cell type–specific transcription factor binding sites could be profiled. Cell type–specific footprints corresponding to more than 100 transcription factors were depleted for both CG and non–CG methyla-tion. Only two factors, CTCF and Zfp410, seemed to lack this correlation between transcription factor binding and methylation level (Mo et al., 2015), which suggests a tight relationship between methylation level and regulatory activity. However, this asso-ciation remains correlational and the causal role of dynamic methylation at these sites remains to be determined (Schübeler, 2015).

In addition to gene expression and transcription factor binding, DNA methylation in both the CG and non–CG contexts can be modulated by the pattern of nucleosome occupancy (Chodavarapu et al., 2010; Lister et al., 2009). In neurons, autocorrelation analysis of non–CG methylation shows a weak modulation with periods of ~10 bp, cor-responding to the DNA helix coil length, and ~180 bp, corresponding to nucleosome spacing (Lister et al., 2013). This phased modulation of DNA methylation is similar in plants and mammals (Chodavarapu et al., 2010), suggesting it is a universal feature of DNA methyltransferase activity. Integrating MethylC-seq data with nucleosome positions estimated from ATAC-seq chromatin accessibility data showed that both CG and non–CG methylation are depleted at the center of nucleosomes and enriched in the space between nucleosomes (Mo et al., 2015).

NON–CG METHYLATION IN X CHROMOSOME INACTIVATION

An intriguing case study of the highly specific, conserved distribution of non–CG methylation across the genome is the X chromosome. In mammals, female cells inactivate one of the two X chromosomes via transcription of the noncoding RNA *Xist* and subsequent epigenetic silencing of transcription from most of the chromosome. X chromosome inactivation is critical for maintaining a balanced dose of X-linked genes (Deng, Berletch, Nguyen, & Disteche, 2014). In neurons and glia, non–CG methylation is lower throughout the X chromosome in females than males, suggesting a differential distribution of this mark on the active versus inactive X (Lister et al., 2013). This female-specific depletion of non–CG methylation contrasts with the established role of CG methylation, which is enriched at the promoters of transcriptionally silenced genes on the inactive X (Sharp et al., 2011).

Although non–CG methylation is depleted across the X chromosome in female cells, it is locally enriched at genes that escape X inactivation and are consequently expressed from both alleles (Lister et al., 2013). This female non–CG methylation signature of escape from X inactivation appears not only in neurons and glia but also in a range of human tissues with much lower overall levels of mCH (Schultz et al., 2015). Non–CG methylation is thus a useful mark for identifying genes with differential epigenetic regulation on the inactive X chromosome. Indeed, analysis of female versus male human tissues identified 109 genes with significantly higher non–CG methylation in females; only nine of these genes showed the female hypermethylation in all tissues, and the remainder were hypermethylated in a tissue-specific manner in one or more tissues (Schultz et al., 2015). This observation also hints at a functional role for non–CG methylation, which could rein in the expression of escape genes that otherwise might experience a two-fold imbalance in transcription between female and male cells (Johnston et al., 2008).

POSSIBLE FUNCTIONS OF NON–CG METHYLATION

The conservation of cell type–specific non–CG methylation patterns between individuals and even across mammalian species, and the strong correlation between non–CG methylation and particular states of transcriptional activity, points to a functional role for this epigenetic mark. To begin to directly address the causal role of non–CG methylation in regulating gene expression, Guo et al. (2014) used an in vitro reporter assay to show that CG methylation (mCG) by the bacterial methyltransferase M.SssI strongly represses transcription; methylation of GC positions by using M.CviPI resulted a similar level of repression. Although GC positions include both CG and non–CG sequences, the majority of such sites are non–CG. This result thus supports a repressive role for non–CG methylation in the regulation of gene expression.

One likely mechanism by which non–CG methylation could act is through the recruitment of methylated DNA binding proteins that reconfigure the local chromatin environment and repress transcription. Methyl-CpG-binding protein 2 (MeCP2) binds methylated CG, CA, and CT sites (Chen et al., 2015; Guo et al., 2014), as well as hmC (Mellén, Ayata, Dewell, Kriaucionis, & Heintz, 2012). Electrophoretic mobility shift assays showed that oligonucleotides containing methylated CA positions compete for MeCP2 binding with equal affinity to oligonucleotides containing methylated CG. Hydroxymethylation at CG positions does not change the affinity for MeCP2 binding, although hmC at CA positions increases the affinity. It seems that hmC is limited to CG positions in vivo (Lister et al., 2013; Yu et al., 2012), suggesting that the major determinants of MeCP2 binding are methylation at CG and CA positions. hmC can function either as a stable base modification or as an intermediate in the TET-mediated active DNA demethylation pathway (Bachman et al., 2014; Guo, Su, Zhong, Ming, & Song, 2011). The finding that hmC does not seem to be present in the non–CG context to any significant level may in part explain why non–CG methylation is so abundant in postmitotic neurons, which at non–CG positions may lack both active (Tet/hmC-mediated) and passive (replication-dependent) DNA demethylation.

The accumulation of non–CG methylation in neurons extends throughout the stages of child and adolescent brain development, spanning the first two decades of life in the human prefrontal cortex (Lister et al., 2013). This period coincides with profound changes to the connectivity of postmitotic neurons, with the mature cortical circuitry emerging through the processes of synaptogenesis and subsequent synaptic pruning. Disruption of brain development during this time can have profound and lasting consequences. It is therefore of considerable interest to determine what role, if any, non–CG methylation plays in directing the proper development of cortical circuitry during early life.

Rett syndrome is a neurodevelopmental disorder that affects girls, with autism-like symptoms emerging around 6–18 months of age at the same time that non–CG methylation begins to accumulate in cortical neurons. Notably, Rett syndrome is caused by mutations in the X-linked gene MeCP2 (Chahrour & Zoghbi, 2007), suggesting a potential causal link between disease onset and disruption of the targeting of MeCP2 to methylated CA sites. Knowledge of the genetic cause of Rett syndrome has enabled development of mouse models of the disease (Ricceri, De Filippis, & Laviola, 2008), which can help define the gene networks that are up- or downregulated by loss of MeCP2 in brain cells.

Notably, genes that are upregulated in MeCP2 knockout animals tend to be long, spanning >100 kb (Gabel et al., 2015). This pattern is observed in a variety of cortical and cerebellar neuron types, including both glutamatergic (excitatory) and GABAergic (inhibitory) cells (Sugino et al., 2014). These long genes, which include many genes with key synaptic function like the calcium/calmodulin-dependent kinase *Camk2d*, are also enriched for gene body CA methylation (Gabel et al., 2015). Moreover, comparing long genes with short genes showing similar levels of CA methylation, MeCP2 seems to selectively repress expression of long genes (Gabel et al., 2015). These data suggest that MeCP2 may repress genes harboring CA methylation in a dose-dependent manner, exerting the strongest influence on long genes with a high total amount of methylcytosine. Furthermore, the developmental timing of the manifestation of Rett syndrome symptoms closely matches the period when non–CG methylation is accumulating in the developing human brain (Chen et al., 2015). This suggests that disruption of MeCP2-dependent recognition of the accumulating non–CG methylation may be linked to the emergence of Rett syndrome.

OUTLOOK AND FUTURE DIRECTIONS

Advances in DNA sequencing technology have enabled a comprehensive assessment of the DNA methylation landscape across the mammalian genome at single-base resolution. Combined with new technology for sorting cell types from a range of different tissues, DNA methylation profiling has revealed a surprising diversity in the sequence context of genomic methylcytosine, including the presence of abundant non–CG methylation. These rapid advances in characterizing the abundance, distribution, and dynamics of non–CG methylation in the genome of mammalian cells have raised a myriad of questions about the form and functional relevance of this distinct type of DNA methylation, which we outline below:

1. **Cell-type specificity**: Despite a number of studies examining non–CG methylation in the mammalian genome, our understanding of the variation in non–CG methylation within and between cell types is very limited. Although improvements have been achieved in resolving the non–CG methylation patterns of a limited number of cell types, most studies have analyzed highly heterogeneous tissues or cellular populations, and we have effectively no understanding of the consistency or variation of non–CG methylation between individual cells. Even in neurons, with the highest concentration of non–CG methylation so far observed, the majority of methylated non–CG sites are only methylated in ~20% of cells (Lister et al., 2013; Schultz et al., 2015). This distribution suggests substantial variability in the non–CG methylation pattern among individual cells. However, recent advances in base resolution DNA methylome profiling from single mammalian cells suggests that this level of analysis may be possible in the near future (Farlik et al., 2015; Smallwood et al., 2014).

2. **Establishment of non–CG methylation**: Although several lines of evidence indicate that DNMT3A establishes non–CG methylation in mature cell types, little is known about how it is recruited to specific regions of the genome to establish this mark. Furthermore, non–CG methylation patterns in a particular tissue type or cell type show very high positional conservation between individuals; however, the cellular factors or genome features that sculpt these precise patterns are unknown. Gaining a better understanding of the mechanistic basis of how the cell targets particular regions of the genome for establishment of, or protection from, non–CG methylation will shed further light on how the observed tissue- or cell-type patterns of non–CG methylation are controlled, enabling subsequent investigation of the potential effects of their disruption upon genome activity.

3. **Information content and function**: Although the presence of non–CG methylation has now been characterized genome-wide in a range of tissues and cell types (Lister et al., 2013, 2009, 2011; Mo et al., 2015; Xie et al., 2012), these studies have not provided direct evidence for a role of non–CG methylation in mediating transcriptional regulation. Recent studies have provided some evidence that disruption of the reading or writing of non–CG methylation in the brain results in altered transcription (Gabel et al., 2015; Guo et al., 2014); however, we have little or no understanding beyond this of the potential function of the modification. Does non–CG methylation only have an effect upon transcription when it is deposited throughout broad regions, such as gene bodies, or can methylation of single non–CG positions affect gene expression? Does deposition of non–CG methylation induce changes in local histone modifications or chromatin state, or is non–CG methylation establishment dependent upon other local histone modifications and chromatin conformation? In-depth investigations are required to explore these questions regarding the potential functionality of non–CG methylation. Investigation of the role of non–CG methylation by inhibition of DNMT3A/B is potentially challenging due to the simultaneous effects upon both CG and non–CG methylation. Alternatively, emerging tools to specifically target changes to DNA methylation at specific locations in the genome may in the future allow direct evaluation of the causal role of non–CG methylation in a range of genomic contexts (Maeder et al., 2013).

4. **Non–CG readers and writers**: An alternative route to investigating the function of non–CG methylation is disruption of the cellular factors that read and interpret the modification, as exemplified by recent studies of MeCP2 (Chen et al., 2015; Gabel et al., 2015). However, beyond MeCP2, the cellular factors that may bind to, read, and interpret the non–CG methylation are unknown. Identification of such factors through unbiased screens would provide new targets for the functional investigation of non–CG methylation and the consequences of disrupting its use in the cell.

5. **Non–CG methylation dynamics**: Although profiling through brain development has revealed the rapid accumulation of non–CG methylation during postnatal maturation of the neural circuitry, little is known about the timing of non–CG methylation accumulation through development in distinct neural cell types and during adult neurogenesis, or potential changes in non–CG methylation in response to neuronal activity and stimulation. Given evidence for the requirement for DNMT3A2 in the adult mammalian hippocampus for cognitive function, and the loss of cognitive capacity upon knockdown or natural age-related cognitive decline of DNMT3A2 abundance in the hippocampus (Oliveira, Hemstedt, & Bading, 2012), an intriguing possibility is that DNMT3A2 is required during neuronal differentiation or maturation during adult neurogenesis. Further investigation of the temporal dynamics of non–CG methylation during neural differentiation and development and in response to neuronal activity and learning is essential to further reveal its functional relevance.

6. **Disruption in neurological disorders**: Recent insights into the relationship between non–CG methylation and MeCP2, the effects of their disruption on upon neural transcription, and the potentially central role that non–CG methylation may play in Rett syndrome, underline the importance of gaining a comprehensive understanding of this modification and the effects of its disruption in human health and disease states. DNA methylation changes at CG positions have been linked with neurodegenerative disorders such as Alzheimer disease (De Jager et al., 2014) as well as with neuropsychiatric disorders such as schizophrenia with a stronger developmental role (Pidsley et al., 2014). Given the dramatic changes in non–CG methylation that take place during human brain development, future comprehensive studies of brain non–CG methylation, and its readers and writers, in neurological disorders of developmental origin need to be undertaken to explore the potential role of disruption of this unique epigenomic mark in human neurological disease.

ACKNOWLEDGMENTS

This work was supported by NIH/NINDS R00NS080911 (EAM), and an ARC Future Fellowship and Sylvia and Charles Viertel Foundation Senior Medical Researh Felloship (RL).

REFERENCES

Bachman, M., et al. (2014). 5-Hydroxymethylcytosine is a predominantly stable DNA modification. *Nature Chemistry, 6*(12), 1049–1055.

Baillie, J. K., Barnett, M. W., Upton, K. R., Gerhardt, D. J., Richmond, T. A., De Sapio, F., et al. (2011). Somatic retrotransposition alters the genetic landscape of the human brain. *Nature, 479*(7374), 534–537. http://doi.org/10.1038/nature10531.

Baubec, T., Colombo, D. F., Wirbelauer, C., Schmidt, J., Burger, L., Krebs, A. R., et al. (2015). Genomic profiling of DNA methyltransferases reveals a role for DNMT3B in genic methylation. *Nature, 520*(7546), 243–247. http://doi.org/10.1038/nature14176.

Berman, B. P., Weisenberger, D. J., Aman, J. F., Hinoue, T., Ramjan, Z., Liu, Y., et al. (2012). Regions of focal DNA hypermethylation and long-range hypomethylation in colorectal cancer coincide with nuclear lamina-associated domains. *Nature Genetics, 44*(1), 40–46. http://doi.org/10.1038/ng.969.

Bourc'his, D., Xu, G.-L., Lin, C.-S., Bollman, B., & Bestor, T. H. (2001). Dnmt3L and the establishment of maternal genomic imprints. *Science (New York, NY), 294*(5551), 2536–2539. http://doi.org/10.1126/science.1065848.

Buenrostro, J. D., Giresi, P. G., Zaba, L. C., Chang, H. Y., & Greenleaf, W. J. (2013). Transposition of native chromatin for fast and sensitive epigenomic profiling of open chromatin, DNA-binding proteins and nucleosome position. *Nature Methods, 10*(12), 1213–1218. http://doi.org/10.1038/nmeth.2688.

Chahrour, M., & Zoghbi, H. Y. (2007). The story of Rett syndrome: from clinic to neurobiology. *Neuron, 56*(3), 422–437. http://doi.org/10.1016/j.neuron.2007.10.001.

Chen, L., Chen, K., Lavery, L. A., Baker, S. A., Shaw, C. A., Li, W., et al. (2015). MeCP2 binds to non-CG methylated DNA as neurons mature, influencing transcription and the timing of onset for Rett syndrome. *Proceedings of the National Academy of Sciences, 112*(17), 5509–5514. http://doi.org/10.1073/pnas.1505909112.

Chodavarapu, R. K., Feng, S., Bernatavichute, Y. V., Chen, P.-Y., Stroud, H., Yu, Y., et al. (2010). Relationship between nucleosome positioning and DNA methylation. *Nature, 466*(7304), 388–392. http://doi.org/10.1038/nature09147.

De Jager, P. L., Srivastava, G., Lunnon, K., Burgess, J., Schalkwyk, L. C., Yu, L., et al. (2014). Alzheimer's disease: early alterations in brain DNA methylation at ANK1, BIN1, RHBDF2 and other loci. *Nature Neuroscience, 17*(9), 1156–1163. http://doi.org/10.1038/nn.3786.

Deng, X., Berletch, J. B., Nguyen, D. K., & Disteche, C. M. (2014). X chromosome regulation: diverse patterns in development, tissues and disease. *Nature Reviews Genetics, 15*(6), 367–378. http://doi.org/10.1038/nrg3687.

Farlik, M., Sheffield, N. C., Nuzzo, A., Datlinger, P., Schönegger, A., Klughammer, J., et al. (2015). Single-cell DNA methylome sequencing and bioinformatic inference of epigenomic cell-state dynamics. *Cell Reports, 10*(8), 1386–1397. http://doi.org/10.1016/j.celrep.2015.02.001.

Gabel, H. W., Kinde, B., Stroud, H., Gilbert, C. S., Harmin, D. A., Kastan, N. R., et al. (2015). Disruption of DNA-methylation-dependent long gene repression in Rett syndrome. *Nature, 522*(7554), 89–93. http://doi.org/10.1038/nature14319.

Guo, J. U., Su, Y., Shin, J. H., Shin, J., Li, H., Xie, B., et al. (2014). Distribution, recognition and regulation of non-CpG methylation in the adult mammalian brain. *Nature Neuroscience, 17*(2), 215–222. http://doi.org/10.1038/nn.3607.

Guo, J. U., Su, Y., Zhong, C., Ming, G. L., & Song, H. (2011). Hydroxylation of 5-methylcytosine by TET1 promotes active DNA demethylation in the adult brain. *Cell, 145*(3), 423–434.

Hon, G. C., Rajagopal, N., Shen, Y., McCleary, D. F., Yue, F., Dang, M. D., et al. (2013). Epigenetic memory at embryonic enhancers identified in DNA methylation maps from adult mouse tissues. *Nature Genetics, 45*(10), 1198–1206. http://doi.org/10.1038/ng.2746.

Johnston, C. M., Lovell, F. L., Leongamornlert, D. A., Stranger, B. E., Dermitzakis, E. T., & Ross, M. T. (2008). Large-scale population study of human cell lines indicates that dosage compensation is virtually complete. *PLoS Genetics, 4*(1), e9. http://doi.org/10.1371/journal.pgen.0040009.st003.

Laurent, L., Wong, E., Li, G., Huynh, T., Tsirigos, A., Ong, C. T., et al. (2010). Dynamic changes in the human methylome during differentiation. *Genome Research, 20*(3), 320–331. http://doi.org/10.1101/gr.101907.109.

Lister, R., & Ecker, J. R. (2009). Finding the fifth base: genome-wide sequencing of cytosine methylation. *Genome Research, 19*(6), 959–966. http://doi.org/10.1101/gr.083451.108.

Lister, R., Mukamel, E. A., Nery, J. R., Urich, M., Puddifoot, C. A., Johnson, N. D., et al. (2013). Global epigenomic reconfiguration during mammalian brain development. *Science (New York, NY), 341*(6146). 1237905 http://doi.org/10.1126/science.1237905.

Lister, R., Pelizzola, M., Dowen, R. H., Hawkins, R. D., Hon, G., Tonti-Filippini, J., et al. (2009). Human DNA methylomes at base resolution show widespread epigenomic differences. *Nature, 462*(7271), 315–322. http://doi.org/10.1038/nature08514.

Lister, R., Pelizzola, M., Kida, Y. S., Hawkins, R. D., Nery, J. R., Hon, G., et al. (2011). Hotspots of aberrant epigenomic reprogramming in human induced pluripotent stem cells. *Nature, 471*(7336), 68–73. http://doi.org/10.1038/nature09798.

Maeder, M. L., Angstman, J. F., Richardson, M. E., Linder, S. J., Cascio, V. M., Tsai, S. Q., et al. (2013). Targeted DNA demethylation and activation of endogenous genes using programmable TALE-TET1 fusion proteins. *Nature Biotechnology, 31*(12), 1137–1142. http://doi.org/10.1038/nbt.2726.

McConnell, M. J., Lindberg, M. R., Brennand, K. J., Piper, J. C., Voet, T., Cowing-Zitron, C., et al. (2013). Mosaic copy number variation in human neurons. *Science (New York, NY), 342*(6158), 632–637. http://doi.org/10.1126/science.1243472.

Mellén, M., Ayata, P., Dewell, S., Kriaucionis, S., & Heintz, N. (2012). MeCP2 binds to 5hmC enriched within active genes and accessible chromatin in the nervous system. *Cell, 151*(7), 1417–1430. http://doi.org/10.1016/j.cell.2012.11.022.

Mo, A., Mukamel, E. A., Davis, F. P., Luo, C., Henry, G. L., Picard, S., et al. (2015). Epigenomic signatures of neuronal diversity in the mammalian brain. *Neuron, 86*(6), 1369–1384. http://doi.org/10.1016/j.neuron.2015.05.018.

Okano, M., Bell, D. W., Haber, D. A., & Li, E. (1999). DNA methyltransferases Dnmt3a and Dnmt3b are essential for de novo methylation and mammalian development. *Cell, 99*(3), 247–257.

Oliveira, A. M. M., Hemstedt, T. J., & Bading, H. (2012). Rescue of aging-associated decline in Dnmt3a2 expression restores cognitive abilities. *Nature Neuroscience, 15*(8), 1111–1113. http://doi.org/10.1038/nn.3151.

Ooi, S. K. T., Qiu, C., Bernstein, E., Li, K., Jia, D., Yang, Z., et al. (2007). DNMT3L connects unmethylated lysine 4 of histone H3 to de novo methylation of DNA. *Nature, 448*(7154), 714–717. http://doi.org/10.1038/nature05987.

Pidsley, R., Viana, J., Hannon, E., Spiers, H., Troakes, C., Al-Saraj, S., et al. (2014). Methylomic profiling of human brain tissue supports a neurodevelopmental origin for schizophrenia. *Genome Biology, 15*(10), 483. http://doi.org/10.1186/s13059-014-0483-2.

Ramsahoye, B. H., Biniszkiewicz, D., Lyko, F., Clark, V., Bird, A. P., & Jaenisch, R. (2000). Non-CpG methylation is prevalent in embryonic stem cells and may be mediated by DNA methyltransferase 3a. *Proceedings of the National Academy of Sciences of the United States of America, 97*(10), 5237–5242.

Ricceri, L., De Filippis, B., & Laviola, G. (2008). Mouse models of Rett syndrome: from behavioural phenotyping to preclinical evaluation of new therapeutic approaches. *Behavioural Pharmacology, 19*(5–6), 501–517. http://doi.org/10.1097/FBP.0b013e32830c3645.

Saxonov, S., Berg, P., & Brutlag, D. L. (2006). A genome-wide analysis of CpG dinucleotides in the human genome distinguishes two distinct classes of promoters. *Proceedings of the National Academy of Sciences of the United States of America, 103*(5), 1412–1417. http://doi.org/10.1073/pnas.0510310103.

Schübeler, D. (2015). Function and information content of DNA methylation. *Nature, 517*(7534), 321–326. http://doi.org/10.1038/nature14192.

Schultz, M. D., He, Y., Whitaker, J. W., Hariharan, M., Mukamel, E. A., Leung, D., et al. (2015). Human body epigenome maps reveal noncanonical DNA methylation variation. *Nature, 523*(7559), 212–216. http://doi.org/10.1038/nature14465.

Sharp, A. J., Stathaki, E., Migliavacca, E., Brahmachary, M., Montgomery, S. B., Dupre, Y., et al. (2011). DNA methylation profiles of human active and inactive X chromosomes. *Genome Research, 21*(10), 1592–1600. http://doi.org/10.1101/gr.112680.110.

Smallwood, S. A., Lee, H. J., Angermueller, C., Krueger, F., Saadeh, H., Peat, J., et al. (2014). Single-cell genome-wide bisulfite sequencing for assessing epigenetic heterogeneity. *Nature Methods, 11*(8), 817–820. http://doi.org/10.1038/nmeth.3035.

Spiegel, I., Mardinly, A. R., Gabel, H. W., Bazinet, J. E., Couch, C. H., Tzeng, C. P., et al. (2014). Npas4 regulates excitatory-inhibitory balance within neural circuits through cell-type-specific gene programs. *Cell, 157*(5), 1216–1229. http://doi.org/10.1016/j.cell.2014.03.058.

Sugino, K., Hempel, C. M., Okaty, B. W., Arnson, H. A., Kato, S., Dani, V. S., et al. (2014). Cell-type-specific repression by methyl-CpG-binding protein 2 is biased toward long genes. *Journal of Neuroscience, 34*(38), 12877–12883. http://doi.org/10.1523/JNEUROSCI.2674-14.2014.

Suzuki, M. M., & Bird, A. (2008). DNA methylation landscapes: provocative insights from epigenomics. *Nature Reviews Genetics, 9*(6), 465–476. http://doi.org/10.1038/nrg2341.

Upton, K. R., Gerhardt, D. J., Jesuadian, J. S., Richardson, S. R., Sánchez-Luque, F. J., Bodea, G. O., et al. (2015). Ubiquitous L1 mosaicism in hippocampal neurons. *Cell, 161*(2), 228–239. http://doi.org/10.1016/j.cell.2015.03.026.

Varley, K. E., Gertz, J., Bowling, K. M., Parker, S. L., Reddy, T. E., Pauli-Behn, F., et al. (2013). Dynamic DNA methylation across diverse human cell lines and tissues. *Genome Research*, *23*(3), 555–567. http://doi.org/10.1101/gr.147942.112.

Xie, W., Barr, C. L., Kim, A., Yue, F., Lee, A. Y., Eubanks, J., et al. (2012). Base-resolution analyses of sequence and parent-of-origin dependent DNA methylation in the mouse genome. *Cell*, *148*(4), 816–831. http://doi.org/10.1016/j.cell.2011.12.035.

Xie, W., Schultz, M. D., Lister, R., Hou, Z., Rajagopal, N., Ray, P., et al. (2013). Epigenomic analysis of multilineage differentiation of human embryonic stem cells. *Cell*, *153*(5), 1134–1148. http://doi.org/10.1016/j.cell.2013.04.022.

Yu, M., Hon, G. C., Szulwach, K. E., Song, C.-X., Zhang, L., Kim, A., et al. (2012). Base-resolution analysis of 5-hydroxymethylcytosine in the mammalian genome. *Cell*, *149*(6), 1368–1380. http://doi.org/10.1016/j.cell.2012.04.027.

Zeng, J., Konopka, G., Hunt, B. G., Preuss, T. M., Geschwind, D., & Yi, S.V. (2012). Divergent whole-genome methylation maps of human and chimpanzee brains reveal epigenetic basis of human regulatory evolution. *American Journal of Human Genetics*, *91*(3), 455–465. http://doi.org/10.1016/j.ajhg.2012.07.024.

Ziller, M. J., Müller, F., Liao, J., Zhang, Y., Gu, H., Bock, C., et al. (2011). Genomic distribution and inter-sample variation of non-CpG methylation across human cell types. *PLoS Genetics*, *7*(12), e1002389. http://doi.org/10.1371/journal.pgen.1002389.

CHAPTER 6

DNA Modifications and Memory

J.J. Day
University of Alabama at Birmingham, Birmingham, AL, United States

INTRODUCTION

The ability to form and store new memory is an essential aspect of neuronal function that is critical for adaptive behavior. One of the cardinal features of behavioral memory is its persistence over days, weeks, months, or even the life span of an organism. This persistence is possible despite the ongoing turnover (degradation and synthesis) of nearly every molecular component in the nervous system, which creates a paradox. How is it that memories can be stored indefinitely in the nervous system despite replacement of nearly every component of the system that was originally modified by the memory? When considered in the context of molecular mechanisms that control neuronal function in memory-related neural circuits, this capacity is not trivial. To understand the neural substrates of learning and memory, we must first understand how single molecules or groups of molecules in the brain can store information across time and perpetuate this information indefinitely into the future after a given experience (Mammen, Huganir, & O'Brien, 1997; Price, Guan, Burlingame, Prusiner, & Ghaemmaghami, 2010). At the molecular level, the perpetuation of a specific activation state that contributes to cellular memory requires that individual molecules have the ability to self-perpetuate a given type of activity despite turnover and in the absence of the stimulus that initiated that activation state. This type of reaction has long been hypothesized to underlie memory (Crick, 1984; Lisman, 1985; Roberson & Sweatt, 2001), making the search for a molecule capable of this type of autoconversion a critical step in elucidating cellular memory storage and cognitive function.

It was based on this realization that Nobel Laureate Francis Crick hypothesized that modification of DNA itself could be a potential mechanism by which information could be stored in the CNS, and propagated beyond waves of molecular turnover (Crick, 1984; Meagher, 2014). Indeed, in early pioneering studies by Boris Vanyushin and colleagues at the Moscow State University, simple forms of learning were shown to induce global changes in DNA methylation in the adult brain (see Chapter 1; Vanyushin, Tushmalova, & Guskova, 1974, 1977). Since that time, the role of individual epigenetic processes in neuronal systems has been the subject of intense exploration. Interestingly, the general form of the chemical reaction, which maintains DNA methylation, is precisely what memory biochemists hypothesized would be necessary for the perpetuation of

DNA Modifications in the Brain
ISBN 978-0-12-801596-4
http://dx.doi.org/10.1016/B978-0-12-801596-4.00006-X

long-term memory traces (Crick, 1984; Lisman, 1985; Razin & Friedman, 1981). Thus, the ability of maintenance DNA methyltransferases (DNMTs) to recognize hemimethylated DNA and remethylate the complementary strand ensures that this modification could remain stable over time despite ongoing molecular turnover within a cell, even in the absence of the initiating agent (Day & Sweatt, 2010). It is now clear that the dynamic regulation of DNA methylation states in neurons is a key contributor to activity-dependent and behavioral processes, most notably in the formation of new memories. This chapter highlights recent research into the cellular and behavioral processes and considers important future questions for the field.

DNA MODIFICATIONS AND NEURONAL MEMORY

The general chemical reaction of DNA methylation suggests that it should be static and long lasting. Direct removal of the methyl moiety from cytosine is extremely difficult given the strength of the carbon–carbon bond. In cases where a methyl group is passively removed from one strand, maintenance DNMT activity (via DNMT1) recognizes hemimethylated DNA, leading again to double-stranded methylation (Ma, Guo, Ming, & Song, 2009). Therefore, this modification is seemingly ideal for the long-term perpetuation of cellular phenotype, gene imprinting, and silencing of repetitive DNA elements (Dulac, 2010; Li, Beard, & Jaenisch, 1993; Reik, Dean, & Walter, 2001). However, the discovery of a role for the Ten-eleven translocation (Tet) family of enzymes in the active demethylation of DNA (via oxidation of methylcytosine to hydroxymethylcytosine) has challenged this canonical view (Guo, Su, Zhong, Ming, & Song, 2011a; Kriaucionis & Heintz, 2009; Tahiliani et al., 2009) and provided a potential mechanism to explain the reversal of established methylation marks in a variety of biological contexts (Day & Sweatt, 2010; Mayer, Niveleau, Walter, Fundele, & Haaf, 2000; Miller & Sweatt, 2007; Pastor et al., 2011; Wu & Zhang, 2010). Importantly, the brain has the highest levels of Tet enzymes of any organ (Kaas et al., 2013; Kriaucionis & Heintz, 2009), suggesting a unique role for DNA demethylation in neuronal function. Indeed, DNA demethylation enzymes have clear roles in embryonic brain development, neuronal differentiation, and neurogenesis (Wu & Zhang, 2010; Xu et al., 2012; Zhang et al., 2013), all of which require a form of cellular "memory" to maintain cellular phenotype across cell division.

DNMTs, TETs, and other DNA methylation–related proteins are also found in relatively high levels in postmitotic, nonproliferating neurons (Dulac, 2010; Goto et al., 1994), suggesting an important functional role in the adult CNS. The most widely accepted candidate for experience-dependent neuronal information storage in the adult brain is synaptic plasticity (Ho, Lee, & Martin, 2011; Malenka & Bear, 2004). This synaptic plasticity can take multiple forms that either adjust the strength of individual synapses, synapses in local dendritic areas, or all synapses across a neuron (Guzman-Karlsson, Meadows, Gavin, Hablitz, & Sweatt, 2014). DNA methylation plays an important role in several different forms of synaptic plasticity. DNMT inhibitors completely block induction of long-term potentiation (LTP) at Schaeffer collateral synapses in the hippocampus, without altering overall synaptic efficacy (Levenson et al., 2006; Miller, Campbell, & Sweatt, 2008). Intriguingly, the effects of DNMT inhibitors are most prominent in late-phase LTP, which is also the component that is sensitive to drugs that inhibit gene transcription and translation (Klann & Dever, 2004; Nguyen, Abel, & Kandel, 1994). Hippocampal LTP is also abolished by conditional deletion of DNMT1 and DNMT3a, whereas long-term depression (LTD) at the same synapses is enhanced after DNMT1/DNMT3a knockout (Feng et al., 2010). Similarly, DNMT activity is required for induction of LTP at cortical and thalamic inputs to the lateral amygdala, suggesting a conserved role for DNMTs in synaptic plasticity across brain structures (Maddox, Watts, & Schafe, 2014; Monsey, Ota, Akingbade, Hong, & Schafe, 2011). Although the role of 5-hydroxymethylcytosine (5hmC) in LTP and LTD remains less clear, one report indicates that knockout of Tet1 results in enhanced hippocampal LTD, without altering induction of LTP (Rudenko et al., 2013).

DNA methylation and hydroxymethylation also have key roles in synaptic scaling, a non-Hebbian form of homeostatic synaptic plasticity that regulates glutamate receptor density in a multiplicative, cell-wide manner. Typically, neuronal silencing leads to synaptic upscaling, whereas prolonged neuronal activation leads to downscaling (Turrigiano, Leslie, Desai, Rutherford, & Nelson, 1998). Similar to LTP, synaptic scaling requires gene transcription and is modulated by DNA methylation at the *Bdnf* gene locus (Nelson, Kavalali, & Monteggia, 2008). Two studies indicate that this synaptic scaling is regulated by DNMT and Tet activity. Pharmacological inhibition or antisense-mediated knockdown of DNMTs results in synaptic upscaling similar to that observed with prolonged neuronal inactivation (Meadows et al., 2015). Similarly, *Tet3* gene expression is bidirectionally regulated by neuronal activity, and overexpression of *Tet3* induces synaptic downscaling. In contrast, knockdown of *Tet3* promotes synaptic upscaling via increases in surface expression of GluR1-containing AMPA receptors (Yu et al., 2015). Together, these results suggest a causal role for DNA methylation and active demethylation in dynamic neuronal responses to environmental stimuli, which is crucial for ongoing neuronal and cognitive function.

DNA MODIFICATION AND BEHAVIORAL MEMORY

Methylation of cytosine bases in DNA has been shown to be essential for a striking variety of memory tasks in numerous species (Day et al., 2013; Feng et al., 2010; Lockett, Helliwell, & Maleszka, 2010; Lubin, Roth, & Sweatt, 2008; Maddox et al., 2014; Miller et al., 2010; Miller & Sweatt, 2007; Monsey et al., 2011; Nikitin, Solntseva, Nikitin, & Kozyrev, 2015). Additionally, emerging evidence has revealed a critical role for Tet enzymes and active DNA demethylation machinery in memory formation (Kaas et al.,

2013; Rudenko et al., 2013). This section reviews the role for DNA methylation and demethylation in different memory systems and highlights the potential involvement of these mechanisms in cognitive disease states.

Spatial and fear memory systems

In rodent model systems, a classic test for memory formation and storage is contextual fear conditioning, in which an animal learns that a novel context (conditioning box) is associated with delivery of a mild aversive electric shock through the floor grid. This learning manifests itself as an increase in the time spent immobile (freezing) upon reexposure to the conditioning chamber, which is extremely robust and can last for the lifetime of an animal. This form of learning is dependent, in part, on the hippocampus, where it requires transcription of genetic information and de novo protein synthesis (Alberini, 2008; Alberini, Milekic, & Tronel, 2006; Davis & Squire, 1984). Early studies investigating the role of DNA methylation in learned behavior used the nucleoside analogue DNMT inhibitors 5-aza-deoxycytidine or zebularine to block DNA methylation in the hippocampus either immediately or 6 h after contextual fear conditioning (Miller & Sweatt, 2007). Strikingly, DNMT inhibitors produced a robust impairment in long-term memory (tested at 24 h after initial training), but *only* when delivered immediately after training. These results are similar to the effects observed with protein synthesis inhibitors and indicated for the first time that DNA methylation was a crucial mechanism in learned behavior. Similarly, infusion of DNMT inhibitors (zebularine or the small molecule inhibitor RG108) directly into the hippocampus before training produced a robust deficit in long-term contextual fear memory (Lubin et al., 2008). Similarly, DNMT inhibition in the lateral amygdala, a brain region critical for cued fear memory, impairs cued fear memory consolidation (Monsey et al., 2011). Importantly, these effects do not seem to be an off-target effect of DNMT inhibitors, as conditional deletion of DNMT1 and DNMT3a in the adult forebrain resulted in impaired contextual fear memory consolidation and impaired spatial memory on the water maze (Feng et al., 2010).

Whereas the initial formation of contextual fear memories is dependent upon the hippocampus, subsequent memory maintenance, remote storage, and retrieval also engages areas of the prefrontal cortex (Bero et al., 2014; Frankland & Bontempi, 2005; Frankland, Bontempi, Talton, Kaczmarek, & Silva, 2004; Lesburgueres et al., 2011; Rajasethupathy et al., 2015; Tse et al., 2011). Consistent with its role in these processes, disruption of the DNA methylation machinery in the cortex has time-dependent effects on memory storage (Miller et al., 2010). Thus, delivery of DNMT inhibitors into the prefrontal cortex shortly after contextual fear conditioning has no effect on subsequent memory maintenance, whereas DNA inhibition in the prefrontal cortex at remote time points after memory formation (eg, 1 month) produces significant degradation of a previously established memory. Similarly, another report demonstrated that long-term object place memory, which is dependent on both the hippocampus and perirhinal cortex, requires DNMT activity in both brain regions (Mitchnick, Creighton, O'Hara, Kalisch, & Winters, 2015). However, further analysis suggested dissociable molecular roles for DNMTs in each brain region. In the hippocampus, knockdown of DNMT3a (but not DNMT1) recapitulated the place learning deficits observed with global DNMT inhibition. Conversely, knockdown of DNMT1 (but not DNMT3a) in the perirhinal cortex impaired long-term memory (Mitchnick et al., 2015). These results suggest that different components of the fear and spatial memory circuitry may engage distinct epigenetic mechanisms at different times (eg, immediately after experiences or at a significant delay) and for different purposes (eg, for recent or remote memory storage).

Consistent with observations that DNA modification is critical for memory formation and storage, several studies have found altered DNA methylation in memory circuits after fear conditioning. For example, formation of contextual fear memories results in hypermethylation of the promoter for the *Ppp1cc* gene, which codes for a subunit of the memory-repressive gene protein phosphatase 1 (Miller & Sweatt, 2007). Conversely, the same experience results in a hypomethylation of the promoter for the memory-promoting gene *Reln* and varied effects on methylation of unique isoform promoters at the *Bdnf* gene locus (Lubin et al., 2008; Miller & Sweatt, 2007; Mizuno, Dempster, Mill, & Giese, 2012). In the hippocampus, these effects seem to be highly dynamic, returning to prememory baseline levels as soon as 24 h after fear conditioning (Miller & Sweatt, 2007). In contrast, cortical changes in DNA methylation have been observed to endure for longer periods, consistent with the behavioral role of cortical structures in memory storage. Thus, *Ppp3ca*, a gene that codes for the catalytic subunit of the memory-suppressing gene calcineurin (a calcium-sensitive protein phosphatase), undergoes promoter hypermethylation and gene repression after fear conditioning, and this change is stable for at least 30 days after memory formation (Miller et al., 2010).

Reports have also revealed a key role for 5hmC in learning and memory. Deletion of *Tet1* results in a significant reduction in 5hmC levels in the cortex and hippocampus, consistent with the involvement of this enzyme in 5-methylcytosine (5mC) hydroxylation (Rudenko et al., 2013). Moreover, *Tet1* knockout mice display impaired extinction of contextual fear and spatial memory, suggestive of an inability adapt to new behavioral contingencies (Rudenko et al., 2013). Conversely, viral overexpression of *Tet1* in the hippocampus increases 5hmC levels and results in impaired long-term contextual fear memory (Kaas et al., 2013). Together, these reports also suggest that Tet1 is a central regulator of immediate early genes such as *Fos*, *Npas4*, *Arc*, and *Egr1*, all of which are induced during memory formation. *Tet1* overexpression results in an upregulation of these key plasticity/memory genes (Kaas et al., 2013), whereas *Tet1* deletion produces a decrease in these transcripts in both hippocampus and cortex (Rudenko et al., 2013).

In contrast to hippocampal *Tet1* manipulations, *Tet1* knockdown in the prefrontal cortex does not alter memory formation or extinction (Li et al., 2014). However, expression of *Tet3*, which is regulated by neuronal activity and extinction training in cortical neurons, is required for normal fear memory extinction. Short hairpin RNA–mediated knockdown of *Tet3* in the infralimbic prefrontal cortex (PFC) results in the maintenance of fear-related freezing behavior despite extinction training (Li et al., 2014). To determine how fear learning and extinction altered 5hmC levels in the PFC, Li and colleagues used 5hmC-immunoprecipitation sequencing, which allowed genome-wide characterization of experience-dependent 5hmC changes. Intriguingly, although the initial fear conditioning did not produce substantial alterations in 5hmC content or localization, extinction training induced profound 5hmC reorganization. Whereas 5hmC was predominantly clustered in intronic and intergenic regions in control animals, extinction training produced a shift in 5hmC peaks in favor of 5′ untranslated regions and coding sequences in DNA (Li et al., 2014). These findings suggest that a "permissive" epigenetic state is established by Tet3 after extinction learning, possibly as a way to establish an epigenetic memory that will alter future experience-dependent gene transcription (Baker-Andresen, Ratnu, & Bredy, 2013; Li, Wei, Ratnu, & Bredy, 2013; Li et al., 2014). Furthermore, these results highlight the partially overlapping but distinct functions of *Tet* family members in the genesis and maintenance of fear-related memories.

Another piece of the DNA methylation/demethylation puzzle is the immediate early gene *Gadd45b*, which despite not being a direct mediator of DNA oxidation, is nevertheless critical for demethylation of cytosine bases in DNA (Guo, Ma, et al., 2011; Ma, Guo, et al., 2009; Ma, Jang, et al., 2009; Niehrs & Schafer, 2012; Sultan & Sweatt, 2013). *Gadd45b* levels are acutely upregulated in the hippocampus and amygdala in mice after exposure to new contexts, including those paired with shock (Leach et al., 2012; Sultan, Wang, Tront, Liebermann, & Sweatt, 2012). However, the role of Gadd45b in memory formation and maintenance is less clear, as different groups have observed opposite effects of Gadd45b deletion on memory capacity (Leach et al., 2012; Sultan et al., 2012). Sultan and colleagues found that *Gadd45b* knockout mice exhibit an increase in long-term fear and spatial memory performance, which is consistent with the enhanced hippocampal LTP observed in these mice (Sultan et al., 2012). In contrast, Leach and colleagues reported a decrease in contextual fear memory in Gadd45b knockout animals (Leach et al., 2012). The cause of this discrepancy is not clear, although the use of different background strains and different training conditions may have influenced these results. Nevertheless, these results support a role for the DNA demethylation enzyme *Gadd45b* in fear memory formation.

Reward-related memory systems

Unlike learned fear responses, the ability to associate predictive environmental cues with reward-related experiences is regulated by the mesolimbic dopamine system. This type of learning can be formally tested by measuring behavioral responses to sensory cues that predict natural rewards (eg, sugar, water). As a result of pairing with rewards, cues themselves begin to elicit anticipatory learned approach behaviors. Dopamine neurons located in the ventral tegmental area (VTA) undergo dynamic alterations in firing rate during the acquisition of these cue–reward associations (Schultz, Dayan, & Montague, 1997), and this process is associated with synaptic plasticity at glutamatergic synapses onto dopamine neurons (Stuber et al., 2008). Furthermore, the activity of dopamine neurons is both necessary and sufficient for learned reward responses (Di Ciano, Cardinal, Cowell, Little, & Everitt, 2001; Tsai et al., 2009). Recently, this form of motivated learning was also shown to depend on DNA methylation in the VTA (Day et al., 2013). Associative cue–reward learning induced a selective upregulation of the immediate early genes *Fos* and *Egr1* in the VTA, and immunohistochemistry using a dopamine cell marker indicated that this increase occurred specifically in dopamine neurons. Moreover, the degree of these changes was correlated with memory acquisition and was associated with altered DNA methylation patterns at these genes, suggesting a potential link between DNA methylation and reward memory formation. Direct infusion of the small molecule DNMT inhibitor RG108 in the VTA before learning produced selective impairment of cue-evoked conditioned responses, without altering reward consumption or baseline behavioral responses. Critically, DNMT inhibition in the VTA after memory acquisition did not impair previously learned behaviors, suggesting that DNA methylation in the VTA is required for the encoding of associative reward memories, but not long-term memory storage or retrieval (Day et al., 2013).

Experience with drugs of abuse also exerts potent control over brain reward circuits and is capable of generating robust memories that can drive addicted individuals to relapse. DNA methylation seems to have a critical role in this process as well (Anier, Malinovskaja, Aonurm-Helm, Zharkovsky, & Kalda, 2010; Feng et al., 2015; LaPlant et al., 2010; Massart et al., 2015). DNMT3a expression is dynamically modulated in the nucleus accumbens (NAc; a key reward structure) after passive or active cocaine administration in animal models, and this change endures even after 28 days of drug withdrawal (LaPlant et al., 2010). Similarly, cocaine experience alters DNA methylation patterns at key genes that have been shown to regulate drug-related behavioral and synaptic responses, such as *Fosb* (Anier et al., 2010). Unlike results in other learning models, blockade of DNMT activity or Cre-mediated DNMT3a excision within the NAc accelerates conditioned place preference memory, whereas herpes simplex virus–mediated overexpression of DNMT3a disrupts cocaine reward memory (LaPlant et al., 2010).

Similarly, expression of *Tet1*, one of the three enzymes responsible for hydroxylation of methylcytosine, is downregulated in the NAc after cocaine experience (Feng et al., 2015). Loss of *Tet1* results in increased cocaine memory, whereas *Tet1* overexpression impairs cocaine place preference. Exposure to cocaine also induces increases in 5hmC levels at numerous genes that are upregulated by cocaine, and knockout of *Tet1* causes increased 5hmC levels at the same loci. This is perhaps counter-intuitive given that *Tet1* has been shown to regulate the conversion of 5mC to 5hmC (Guo, Su, Zhong, Ming, &

Song, 2011b; Kaas et al., 2013; Tahiliani et al., 2009). However, given that *Tet1* may also be involved in the generation of further oxidative species 5-formylcytosine (5fC) and 5-carboxylcytosine, it is possible that 5hmC accumulation after *Tet1* loss is the result of the lack of this continued oxidation (Raiber et al., 2015) or potentially through other noncatalytic roles of Tet1 in the regulation of gene expression (Tsai et al., 2014). Nevertheless, it is clear that DNA methylation and hydroxymethylation play a crucial role in drug-related behavioral memory, indicating that selective targeting of DNA methylation processes may be a potential avenue to targeted addiction therapeutics.

DNA methylation in brain diseases involving cognitive deficits

Finally, although this topic has been reviewed extensively previously (Day, Kennedy, & Sweatt, 2015; Day & Sweatt, 2012; Graff & Tsai, 2013), there is a strong link between DNA modifications and cognitive disease states. In addition to the classical case of Rett syndrome (a neurodevelopmental disability caused by mutations in the X-linked DNA methylation reader methyl-CpG-binding protein 2) (Amir et al., 1999), alterations in DNA methylation pathways have been implicated in Fragile X syndrome (Sutcliffe et al., 1992), Alzheimer disease (Coppieters et al., 2013; De Jager et al., 2014), frontotemporal dementia (Liu et al., 2014), and age-associated cognitive decline (Oliveira, Hemstedt, & Bading, 2012). In some cases, DNA methylation states can serve as nongenetic markers of disease state or prognosis, enabling potential use as disease biomarkers (De Jager et al., 2014; Liu et al., 2014). In addition, there is growing interest in the potential use of DNA methylation–targeted therapeutics for cognitive disease states (Day et al., 2015; Day & Roberson, 2015), although much further research and exploration are required to translate our current understanding of DNA modifications in memory to clinical settings.

FUTURE DIRECTIONS

How is DNA methylation regulated at specific genes?

Neuronal activity, brain development, memory formation, and cognitive disease states are all associated with reorganization of DNA methylation patterns at specific genes with defined temporal dynamics (Day et al., 2013; De Jager et al., 2014; Guo, Ma, et al., 2011; Lim et al., 2014; Lister et al., 2013; Lubin et al., 2008; Miller et al., 2010; Miller & Sweatt, 2007; Nwaobi, Lin, Peramsetty, & Olsen, 2014). This observation suggests active regulation of methylation and demethylation targeting at specific genes and possibly even specific cytosine nucleotides. Although some DNA methylation patterns could be attributed to the local sequence preferences of DNMT1 and DNMT3a (Handa & Jeltsch, 2005) or to the formation of complexes with other DNA binding proteins (Robertson et al., 2000), this does not explain how specific genes undergo experience-dependent changes in DNA methylation. In some cases, behavioral experiences can directly alter the expression of DNA methylation machinery, as observed for *Dnmt3a* (Miller & Sweatt, 2007), *Gadd45b* (Sultan et al., 2012), *Tet1* (Kaas et al., 2013), and *Tet3* (Li et al., 2014). However, these expression changes would result in global alterations in levels of these enzymes, making it unclear how epigenetic specificity can be obtained.

Noncoding RNAs represent one possible gene-specific regulator of DNA modification processes. For example, piwi-interacting RNAs (piRNAs) are a short (26–31 nucleotide) noncoding RNA species that are important for methylation-induced silencing of transposable elements in the germline (Aravin, Sachidanandam, Girard, Fejes-Toth, & Hannon, 2007; Brennecke et al., 2008). piRNAs were also found to exist in the *Aplysia* CNS, where they are increased in response to a serotonin stimulation protocol that induces long-term synaptic facilitation (Rajasethupathy et al., 2012). Although the precise mechanism is not known, piRNA induction was found to regulate CREB2 promoter methylation via DNMT activity, resulting in a sustained promoter hypermethylation after serotonin stimulation (Rajasethupathy et al., 2012). Similarly, long noncoding RNAs have been implicated in direct control over DNA methylation via interactions with DNA methyltransferases (Di Ruscio et al., 2013; Holz-Schietinger & Reich, 2012). In contrast to piRNAs, which are associated with gene methylation, these long noncoding RNAs have an inhibitory relationship with DNA methylation, likely by binding to DNMTs and blocking catalytic activity. Importantly, one class of DNMT-interacting RNAs (termed extra-coding RNA) are synthesized from genomic loci that overlap protein coding genes, in effect establishing a gene-specific mechanism for control of methylation status at that gene (Di Ruscio et al., 2013). However, neither piRNAs nor DNMT-interacting long noncoding RNAs have been investigated in the vertebrate nervous system, making it unclear how these mechanisms may contribute to neuronal plasticity and memory formation. Additional studies are required to understand how behavioral experiences result in gene-specific DNA modifications, and whether specific modifications regulate memory strength and persistence.

How does DNA methylation regulate cell-wide or synapse-specific plasticity to control memory?

Together, the evidence already mentioned highlights a central role for DNA methylation and demethylation in experience-dependent plasticity of fear-related brain circuits. However, although the same processes have been shown to regulate synaptic and homeostatic plasticity, it remains unclear how DNA modifications might affect the actual output of neurons within these defined brain circuits to alter memory function. Neurons in the dorsal CA1 of the hippocampus exhibit patterned activation in specific spatial locations of an environment, providing individual elements of a spatial map that is hypothesized to be central to the role of the hippocampus in spatial memories (Moser, Kropff, & Moser, 2008; O'Keefe & Dostrovsky, 1971).

Presumably, DNA modifications regulate this capability by influencing transcriptional programs that control the plasticity, stability, or both of place fields of hippocampal neurons (Fig. 6.1). This regulation could potentially disrupt or augment the informational content encoded by CA1 neurons during spatial navigation, which would read out at the behavioral level as impaired or enhanced memory. For example, one study used in vivo neurophysiological recordings to examine place cell stability in rats performing a foraging task in a circular track (Roth et al., 2015). Typically, cells in the CA1 and CA3 of the hippocampus maintain the same place fields upon multiple exposures to given environment. However, animals that received intrahippocampal infusions of the DNMT inhibitor zebularine exhibited reduced place field stability between exposures to a novel environment compared to animals receiving a control vehicle injection (Roth et al., 2015). These results suggest a link between DNA methylation and the strength of place cell maps in the hippocampus, both of which are necessary for memory persistence. More work is required to dissect the specificity of this process and to understand how DNA methylation contributes to neuronal output in brain areas that encode behavioral events differently.

How do other DNA modifications contribute to memory formation?

With the discovery that Tet proteins contribute not only to the conversion of 5mC to 5hmC but also to the creation of oxidized derivatives 5fC and 5-carboxylcytosine (Ito et al., 2011), an important topic for additional research will be to investigate the potential role of these modifications in learning and memory. These bases can be excised and removed from DNA via the thymine–DNA glycosylase pathway, giving rise to the hypothesis that these modifications are temporary intermediates leading to DNA demethylation (Maiti & Drohat, 2011). However, evidence indicates that an abundance of transcription factors and chromatin regulators interact directly with 5fC, demonstrating a possible regulatory role of this base modification (Iurlaro et al., 2013). 5fC is also developmentally regulated, is enriched in brain tissue, can be highly stable in vivo (Bachman et al., 2015), and maps to enhancer sites known to modulate gene expression (Song et al., 2013). These results highlight the potential relevance of 5hmC derivatives (beyond serving as intermediates in DNA demethylation), which will be assisted by the genome-wide profiling techniques described in Chapter 2 (Plongthongkum, Diep, & Zhang, 2014; Song et al., 2013).

CONCLUSIONS

Although the investigation of epigenetic mechanisms in the brain remains in its early stages, the results outlined herein reveal an essential role for DNA modifications in several molecular, synaptic, and behavioral processes that regulate memory formation (see

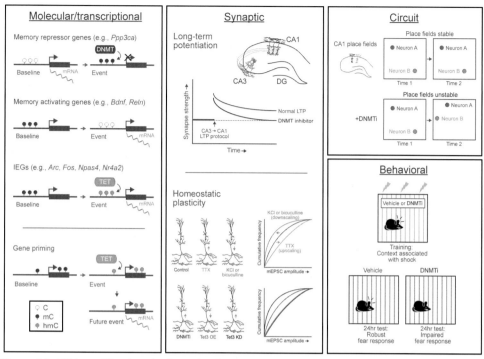

Figure 6.1 *DNA modifications in the molecular/genetic, synaptic, and systems alterations that regulate memory formation.* (left panel) Transcriptional programs regulated by DNA modification include methylation of memory repressor genes and demethylation of memory activator genes during a learning event. Immediate early genes (IEGs) are also induced during memory formation, and this process is impaired in animals lacking *Tet1*. A second, but not mutually exclusive mechanism is gene priming, wherein 5-hydroxymethylcytosine (5hmC) peaks in DNA are remapped to gene bodies and promoters and result in an altered response to future events, but no baseline change in gene expression after the initial event. Key: C, cytosine; mC, methylcytosine. (middle panel) DNA modifications in synaptic plasticity. Treatment with DNA methyltransferases (DNMT) inhibitors before induction of long-term potentiation (LTP) in hippocampal slices blocks late-phase LTP at Schaeffer collaterals (CA3→CA1 synapses). Homeostatic plasticity in response to inactivation (with tetrodotoxin, TTX) or prolonged stimulation (with KCl or bicuculline) results in synaptic upscaling or downscaling, respectively. This is observed as increases (upscaling) or decreases (downscaling) in miniexcitatory postsynaptic current (mEPSC) amplitudes in cultured neurons. DNMT inhibitors and knockdown (KD) of *Tet3*, both of which increase DNA methylation at activity-regulated genes such as Bdnf, results in upscaling. Overexpression of *Tet3* induces synaptic downscaling. (right top panel) Circuit effects of DNA modifications. Hippocampal place fields are normally stable between exposures to a novel environment, but are less stable when a DNMT inhibitor (DNMTi) is infused into the hippocampus. (bottom right panel) DNA modifications in behavioral memory. Consistent with transcriptional, synaptic, and circuit effects of DNMT inhibition, blocking DNMT activity during memory formation results in impaired long-term contextual fear memory.

Fig. 6.1). DNA methylation and demethylation actively regulate gene transcription and the transcriptional potential of the genome with precise temporal and cellular dynamics after behavioral experiences. The same processes are critical for synaptic alterations including LTP and synaptic scaling, two of the leading candidates for long-lasting changes in neuronal activity. Similarly, DNA modifications have been implicated in both memory formation and long-term storage in several brain areas, suggesting convergent molecular regulation of distinct behavioral processes. Together with evidence highlighted in other chapters of this book, these observations have broad implications for our understanding of the molecular processes that regulate normal and pathological function in the brain.

ACKNOWLEDGMENTS

JJD is supported by the National Institute on Drug Abuse (DA034681 & DA039650), startup funds from University of Alabama, and the Evelyn F. McKnight Brain Research Foundation.

REFERENCES

Alberini, C. M. (2008). The role of protein synthesis during the labile phases of memory: revisiting the skepticism. *Neurobiology of Learning and Memory, 89*(3), 234–246.

Alberini, C. M., Milekic, M. H., & Tronel, S. (2006). Mechanisms of memory stabilization and de-stabilization. *Cellular and Molecular Life Sciences: CMLS, 63*(9), 999–1008.

Amir, R. E., Van den Veyver, I. B., Wan, M., Tran, C. Q., Francke, U., & Zoghbi, H. Y. (1999). Rett syndrome is caused by mutations in X-linked MECP2, encoding methyl-CpG-binding protein 2. *Nature Genetics, 23*(2), 185–188.

Anier, K., Malinovskaja, K., Aonurm-Helm, A., Zharkovsky, A., & Kalda, A. (2010). Dna methylation regulates cocaine-induced behavioral sensitization in mice. *Neuropsychopharmacology: Official Publication of the American College of Neuropsychopharmacology, 35*(12), 2450–2461.

Aravin, A. A., Sachidanandam, R., Girard, A., Fejes-Toth, K., & Hannon, G. J. (2007). Developmentally regulated piRNA clusters implicate MILI in transposon control. *Science, 316*(5825), 744–747. http://dx.doi.org/10.1126/science.1142612.

Bachman, M., Uribe-Lewis, S., Yang, X., Burgess, H. E., Iurlaro, M., Reik, W., et al. (2015). 5-Formylcytosine can be a stable DNA modification in mammals. *Nature Chemical Biology.* http://dx.doi.org/10.1038/nchembio.1848.

Baker-Andresen, D., Ratnu, V. S., & Bredy, T. W. (2013). Dynamic DNA methylation: a prime candidate for genomic metaplasticity and behavioral adaptation. *Trends in Neurosciences, 36*(1), 3–13. http://dx.doi.org/10.1016/j.tins.2012.09.003.

Bero, A. W., Meng, J., Cho, S., Shen, A. H., Canter, R. G., Ericsson, M., et al. (2014). Early remodeling of the neocortex upon episodic memory encoding. *Proceedings of the National Academy of Sciences USA, 111*(32), 11852–11857. http://dx.doi.org/10.1073/pnas.1408378111.

Brennecke, J., Malone, C. D., Aravin, A. A., Sachidanandam, R., Stark, A., & Hannon, G. J. (2008). An epigenetic role for maternally inherited piRNAs in transposon silencing. *Science, 322*(5906), 1387–1392. http://dx.doi.org/10.1126/science.1165171.

Coppieters, N., Dieriks, B. V., Lill, C., Faull, R. L., Curtis, M. A., & Dragunow, M. (2013). Global changes in DNA methylation and hydroxymethylation in Alzheimer's disease human brain. *Neurobiology of Aging.* http://dx.doi.org/10.1016/j.neurobiolaging.2013.11.031.

Crick, F. (1984). Memory and molecular turnover. *Nature, 312*(5990), 101.

Davis, H. P., & Squire, L. R. (1984). Protein synthesis and memory: a review. *Psychological Bulletin, 96*(3), 518–559.

Day, J. J., Childs, D., Guzman-Karlsson, M. C., Kibe, M., Moulden, J., Song, E., et al. (2013). DNA methylation regulates associative reward learning. *Nature Neuroscience, 16*(10), 1445–1452. http://dx.doi.org/10.1038/nn.3504.

Day, J. J., Kennedy, A. J., & Sweatt, J. D. (2015). DNA methylation and its implications and accessibility for neuropsychiatric therapeutics. *Annual Review of Pharmacology and Toxicology, 55,* 591–611. http://dx.doi.org/10.1146/annurev-pharmtox-010814-124527.

Day, J. J., & Roberson, E. D. (2015). DNA methylation slows effects of C9orf72 mutations: an epigenetic brake on genetic inheritance. *Neurology, 84*(16), 1616–1617. http://dx.doi.org/10.1212/WNL.0000000000001504.

Day, J. J., & Sweatt, J. D. (2010). DNA methylation and memory formation. *Nature Neuroscience, 13*(11), 1319–1323. http://dx.doi.org/10.1038/nn.2666.

Day, J. J., & Sweatt, J. D. (2012). Epigenetic treatments for cognitive impairments. *Neuropsychopharmacology: Official Publication of the American College of Neuropsychopharmacology, 37*(1), 247–260. http://dx.doi.org/10.1038/npp.2011.85.

De Jager, P. L., Srivastava, G., Lunnon, K., Burgess, J., Schalkwyk, L. C., Yu, L., et al. (2014). Alzheimer's disease: early alterations in brain DNA methylation at ANK1, BIN1, RHBDF2 and other loci. *Nature Neuroscience, 17*(9), 1156–1163. http://dx.doi.org/10.1038/nn.3786.

Di Ciano, P., Cardinal, R. N., Cowell, R. A., Little, S. J., & Everitt, B. J. (2001). Differential involvement of NMDA, AMPA/kainate, and dopamine receptors in the nucleus accumbens core in the acquisition and performance of pavlovian approach behavior. *The Journal of Neuroscience: The Official Journal of the Society for Neuroscience, 21*(23), 9471–9477.

Di Ruscio, A., Ebralidze, A. K., Benoukraf, T., Amabile, G., Goff, L. A., Terragni, J., et al. (2013). DNMT1-interacting RNAs block gene-specific DNA methylation. *Nature, 503*(7476), 371–376. http://dx.doi.org/10.1038/nature12598.

Dulac, C. (2010). Brain function and chromatin plasticity. *Nature, 465*(7299), 728–735.

Feng, J., Shao, N., Szulwach, K. E., Vialou, V., Huynh, J., Zhong, C., et al. (2015). Role of Tet1 and 5-hydroxymethylcytosine in cocaine action. *Nature Neuroscience, 18*(4), 536–544. http://dx.doi.org/10.1038/nn.3976.

Feng, J., Zhou, Y., Campbell, S. L., Le, T., Li, E., Sweatt, J. D., et al. (2010). Dnmt1 and Dnmt3a maintain DNA methylation and regulate synaptic function in adult forebrain neurons. *Nature Neuroscience, 13*(4), 423–430.

Frankland, P. W., & Bontempi, B. (2005). The organization of recent and remote memories. *Nature Reviews. Neuroscience, 6*(2), 119–130.

Frankland, P. W., Bontempi, B., Talton, L. E., Kaczmarek, L., & Silva, A. J. (2004). The involvement of the anterior cingulate cortex in remote contextual fear memory. *Science, 304*(5672), 881–883.

Goto, K., Numata, M., Komura, J. I., Ono, T., Bestor, T. H., & Kondo, H. (1994). Expression of DNA methyltransferase gene in mature and immature neurons as well as proliferating cells in mice. *Differentiation; Research in Biological Diversity, 56*(1–2), 39–44.

Graff, J., & Tsai, L. H. (2013). Histone acetylation: molecular mnemonics on the chromatin. *Nature Reviews. Neuroscience, 14*(2), 97–111. http://dx.doi.org/10.1038/nrn3427.

Guo, J. U., Ma, D. K., Mo, H., Ball, M. P., Jang, M. H., Bonaguidi, M. A., et al. (2011). Neuronal activity modifies the DNA methylation landscape in the adult brain. *Nature Neuroscience, 14*(10), 1345–1351.

Guo, J. U., Su, Y., Zhong, C., Ming, G. L., & Song, H. (2011a). Emerging roles of TET proteins and 5-hydroxymethylcytosines in active DNA demethylation and beyond. *Cell Cycle, 10*(16), 2662–2668.

Guo, J. U., Su, Y., Zhong, C., Ming, G. L., & Song, H. (2011b). Hydroxylation of 5-methylcytosine by TET1 promotes active DNA demethylation in the adult brain. *Cell, 145*(3), 423–434.

Guzman-Karlsson, M. C., Meadows, J. P., Gavin, C. F., Hablitz, J. J., & Sweatt, J. D. (2014). Transcriptional and epigenetic regulation of Hebbian and non-Hebbian plasticity. *Neuropharmacology, 80,* 3–17. http://dx.doi.org/10.1016/j.neuropharm.2014.01.001.

Handa, V., & Jeltsch, A. (2005). Profound flanking sequence preference of Dnmt3a and Dnmt3b mammalian DNA methyltransferases shape the human epigenome. *Journal of Molecular Biology, 348*(5), 1103–1112. http://dx.doi.org/10.1016/j.jmb.2005.02.044.

Ho, V. M., Lee, J. A., & Martin, K. C. (2011). The cell biology of synaptic plasticity. *Science, 334*(6056), 623–628. http://dx.doi.org/10.1126/science.1209236.

Holz-Schietinger, C., & Reich, N. O. (2012). RNA modulation of the human DNA methyltransferase 3A. *Nucleic Acids Research, 40*(17), 8550–8557. http://dx.doi.org/10.1093/nar/gks537.

Ito, S., Shen, L., Dai, Q., Wu, S. C., Collins, L. B., Swenberg, J. A., et al. (2011). Tet proteins can convert 5-methylcytosine to 5-formylcytosine and 5-carboxylcytosine. *Science, 333*(6047), 1300–1303. http://dx.doi.org/10.1126/science.1210597.

Iurlaro, M., Ficz, G., Oxley, D., Raiber, E. A., Bachman, M., Booth, M. J., et al. (2013). A screen for hydroxy-methylcytosine and formylcytosine binding proteins suggests functions in transcription and chromatin regulation. *Genome Biology, 14*(10), R119. http://dx.doi.org/10.1186/gb-2013-14-10-r119.

Kaas, G. A., Zhong, C., Eason, D. E., Ross, D. L., Vachhani, R. V., Ming, G. L., et al. (2013). TET1 controls CNS 5-methylcytosine hydroxylation, active DNA demethylation, gene transcription, and memory formation. *Neuron, 79*(6), 1086–1093. http://dx.doi.org/10.1016/j.neuron.2013.08.032.

Klann, E., & Dever, T. E. (2004). Biochemical mechanisms for translational regulation in synaptic plasticity. *Nature Reviews. Neuroscience, 5*(12), 931–942. http://dx.doi.org/10.1038/nrn1557.

Kriaucionis, S., & Heintz, N. (2009). The nuclear DNA base 5-hydroxymethylcytosine is present in Purkinje neurons and the brain. *Science, 324*(5929), 929–930.

LaPlant, Q., Vialou, V., Covington, H. E., 3rd, Dumitriu, D., Feng, J., Warren, B. L., et al. (2010). Dnmt3a regulates emotional behavior and spine plasticity in the nucleus accumbens. *Nature Neuroscience, 13*(9), 1137–1143.

Leach, P. T., Poplawski, S. G., Kenney, J. W., Hoffman, B., Liebermann, D. A., Abel, T., et al. (2012). Gadd45b knockout mice exhibit selective deficits in hippocampus-dependent long-term memory. *Learning & Memory, 19*(8), 319–324. http://dx.doi.org/10.1101/lm.024984.111.

Lesburgueres, E., Gobbo, O. L., Alaux-Cantin, S., Hambucken, A., Trifilieff, P., & Bontempi, B. (2011). Early tagging of cortical networks is required for the formation of enduring associative memory. *Science, 331*(6019), 924–928. http://dx.doi.org/10.1126/science.1196164.

Levenson, J. M., Roth, T. L., Lubin, F. D., Miller, C. A., Huang, I. C., Desai, P., et al. (2006). Evidence that DNA (cytosine-5) methyltransferase regulates synaptic plasticity in the hippocampus. *The Journal of Biological Chemistry, 281*(23), 15763–15773. http://dx.doi.org/10.1074/jbc.M511767200.

Li, E., Beard, C., & Jaenisch, R. (1993). Role for DNA methylation in genomic imprinting. *Nature, 366*(6453), 362–365. http://dx.doi.org/10.1038/366362a0.

Lim, A. S., Srivastava, G. P., Yu, L., Chibnik, L. B., Xu, J., Buchman, A. S., et al. (2014). 24-hour rhythms of DNA methylation and their relation with rhythms of RNA expression in the human dorsolateral prefrontal cortex. *PLoS Genetics, 10*(11), e1004792. http://dx.doi.org/10.1371/journal.pgen.1004792.

Lisman, J. E. (1985). A mechanism for memory storage insensitive to molecular turnover: a bistable auto-phosphorylating kinase. *Proceedings of the National Academy of Sciences USA, 82*(9), 3055–3057.

Lister, R., Mukamel, E. A., Nery, J. R., Urich, M., Puddifoot, C. A., Johnson, N. D., et al. (2013). Global epigenomic reconfiguration during mammalian brain development. *Science, 341*(6146), 1237905. http://dx.doi.org/10.1126/science.1237905.

Liu, E. Y., Russ, J., Wu, K., Neal, D., Suh, E., McNally, A. G., et al. (2014). *C9orf72* hypermethylation protects against repeat expansion-associated pathology in ALS/FTD. *Acta Neuropathologica, 128*(4), 525–541. http://dx.doi.org/10.1007/s00401-014-1286-y.

Li, X., Wei, W., Ratnu, V. S., & Bredy, T. W. (2013). On the potential role of active DNA demethylation in establishing epigenetic states associated with neural plasticity and memory. *Neurobiology of Learning and Memory, 105*, 125–132. http://dx.doi.org/10.1016/j.nlm.2013.06.009.

Li, X., Wei, W., Zhao, Q. Y., Widagdo, J., Baker-Andresen, D., Flavell, C. R., et al. (2014). Neocortical Tet3-mediated accumulation of 5-hydroxymethylcytosine promotes rapid behavioral adaptation. *Proceedings of the National Academy of Sciences USA, 111*(19), 7120–7125. http://dx.doi.org/10.1073/pnas.1318906111.

Lockett, G. A., Helliwell, P., & Maleszka, R. (2010). Involvement of DNA methylation in memory processing in the honey bee. *Neuroreport, 21*(12), 812–816. http://dx.doi.org/10.1097/WNR.0b013e32833ce5be.

Lubin, F. D., Roth, T. L., & Sweatt, J. D. (2008). Epigenetic regulation of BDNF gene transcription in the consolidation of fear memory. *The Journal of Neuroscience: The Official Journal of the Society for Neuroscience, 28*(42), 10576–10586.

Maddox, S. A., Watts, C. S., & Schafe, G. E. (2014). DNA methyltransferase activity is required for memory-related neural plasticity in the lateral amygdala. *Neurobiology of Learning and Memory, 107*, 93–100. http://dx.doi.org/10.1016/j.nlm.2013.11.008.

Ma, D. K., Guo, J. U., Ming, G. L., & Song, H. (2009). DNA excision repair proteins and Gadd45 as molecular players for active DNA demethylation. *Cell Cycle, 8*(10), 1526–1531.

Maiti, A., & Drohat, A. C. (2011). Thymine DNA glycosylase can rapidly excise 5-formylcytosine and 5-carboxylcytosine: potential implications for active demethylation of CpG sites. *The Journal of Biological Chemistry*, *286*(41), 35334–35338. http://dx.doi.org/10.1074/jbc.C111.284620.

Ma, D. K., Jang, M. H., Guo, J. U., Kitabatake, Y., Chang, M. L., Pow-Anpongkul, N., et al. (2009). Neuronal activity-induced Gadd45b promotes epigenetic DNA demethylation and adult neurogenesis. *Science*, *323*(5917), 1074–1077.

Malenka, R. C., & Bear, M. F. (2004). LTP and LTD: an embarrassment of riches. *Neuron*, *44*(1), 5–21. http://dx.doi.org/10.1016/j.neuron.2004.09.012.

Mammen, A. L., Huganir, R. L., & O'Brien, R. J. (1997). Redistribution and stabilization of cell surface glutamate receptors during synapse formation. *The Journal of Neuroscience: the Official Journal of the Society for Neuroscience*, *17*(19), 7351–7358.

Massart, R., Barnea, R., Dikshtein, Y., Suderman, M., Meir, O., Hallett, M., et al. (2015). Role of DNA methylation in the nucleus accumbens in incubation of cocaine craving. *The Journal of Neuroscience: The Official Journal of the Society for Neuroscience*, *35*(21), 8042–8058. http://dx.doi.org/10.1523/JNEUROSCI.3053-14.2015.

Mayer, W., Niveleau, A., Walter, J., Fundele, R., & Haaf, T. (2000). Demethylation of the zygotic paternal genome. *Nature*, *403*(6769), 501–502. http://dx.doi.org/10.1038/35000654.

Meadows, J. P., Guzman-Karlsson, M. C., Phillips, S., Holleman, C., Posey, J. L., Day, J. J., et al. (2015). DNA methylation regulates neuronal glutamatergic synaptic scaling. *Science Signaling*, *8*(382), 61. http://dx.doi.org/10.1126/scisignal.aab0715.

Meagher, R. B. (2014). 'Memory and molecular turnover,' 30 years after inception. *Epigenetics Chromatin*, *7*(1), 37. http://dx.doi.org/10.1186/1756-8935-7-37.

Miller, C. A., Campbell, S. L., & Sweatt, J. D. (2008). DNA methylation and histone acetylation work in concert to regulate memory formation and synaptic plasticity. *Neurobiology of Learning and Memory*, *89*(4), 599–603. http://dx.doi.org/10.1016/j.nlm.2007.07.016.

Miller, C. A., Gavin, C. F., White, J. A., Parrish, R. R., Honasoge, A., Yancey, C. R., et al. (2010). Cortical DNA methylation maintains remote memory. *Nature Neuroscience*, *13*(6), 664–666. http://dx.doi.org/10.1038/nn.2560.

Miller, C. A., & Sweatt, J. D. (2007). Covalent modification of DNA regulates memory formation. *Neuron*, *53*(6), 857–869. http://dx.doi.org/10.1016/j.neuron.2007.02.022.

Mitchnick, K. A., Creighton, S., O'Hara, M., Kalisch, B. E., & Winters, B. D. (2015). Differential contributions of de novo and maintenance DNA methyltransferases to object memory processing in the rat hippocampus and perirhinal cortex – a double dissociation. *The European Journal of Neuroscience*, *41*(6), 773–786. http://dx.doi.org/10.1111/ejn.12819.

Mizuno, K., Dempster, E., Mill, J., & Giese, K. P. (2012). Long-lasting regulation of hippocampal Bdnf gene transcription after contextual fear conditioning. *Genes, Brain, and Behavior*, *11*(6), 651–659. http://dx.doi.org/10.1111/j.1601-183X.2012.00805.x.

Monsey, M. S., Ota, K. T., Akingbade, I. F., Hong, E. S., & Schafe, G. E. (2011). Epigenetic alterations are critical for fear memory consolidation and synaptic plasticity in the lateral amygdala. *PLoS One*, *6*(5), e19958. http://dx.doi.org/10.1371/journal.pone.0019958.

Moser, E. I., Kropff, E., & Moser, M. B. (2008). Place cells, grid cells, and the brain's spatial representation system. *Annual Review of Neuroscience*, *31*, 69–89. http://dx.doi.org/10.1146/annurev.neuro.31.061307.090723.

Nelson, E. D., Kavalali, E. T., & Monteggia, L. M. (2008). Activity-dependent suppression of miniature neurotransmission through the regulation of DNA methylation. *The Journal of Neuroscience: The Official Journal of the Society for Neuroscience*, *28*(2), 395–406. http://dx.doi.org/10.1523/JNEUROSCI.3796-07.2008.

Nguyen, P. V., Abel, T., & Kandel, E. R. (1994). Requirement of a critical period of transcription for induction of a late phase of LTP. *Science*, *265*(5175), 1104–1107.

Niehrs, C., & Schafer, A. (2012). Active DNA demethylation by Gadd45 and DNA repair. *Trends in Cell Biology*, *22*(4), 220–227. http://dx.doi.org/10.1016/j.tcb.2012.01.002.

Nikitin, V. P., Solntseva, S. V., Nikitin, P. V., & Kozyrev, S. A. (2015). The role of DNA methylation in the mechanisms of memory reconsolidation and development of amnesia. *Behavioural Brain Research*, *279*, 148–154. http://dx.doi.org/10.1016/j.bbr.2014.11.025.

Nwaobi, S. E., Lin, E., Peramsetty, S. R., & Olsen, M. L. (2014). DNA methylation functions as a critical regulator of Kir4.1 expression during CNS development. *Glia, 62*(3), 411–427. http://dx.doi.org/10.1002/glia.22613.

O'Keefe, J., & Dostrovsky, J. (1971). The hippocampus as a spatial map. Preliminary evidence from unit activity in the freely-moving rat. *Brain Research, 34*(1), 171–175.

Oliveira, A. M., Hemstedt, T. J., & Bading, H. (2012). Rescue of aging-associated decline in Dnmt3a2 expression restores cognitive abilities. *Nature Neuroscience, 15*(8), 1111–1113.

Pastor, W. A., Pape, U. J., Huang, Y., Henderson, H. R., Lister, R., Ko, M., et al. (2011). Genome-wide mapping of 5-hydroxymethylcytosine in embryonic stem cells. *Nature, 473*(7347), 394–397. http://dx.doi.org/10.1038/nature10102.

Plongthongkum, N., Diep, D. H., & Zhang, K. (2014). Advances in the profiling of DNA modifications: cytosine methylation and beyond. *Nature Reviews. Genetics, 15*(10), 647–661. http://dx.doi.org/10.1038/nrg3772.

Price, J. C., Guan, S., Burlingame, A., Prusiner, S. B., & Ghaemmaghami, S. (2010). Analysis of proteome dynamics in the mouse brain. *Proceedings of the National Academy of Sciences USA, 107*(32), 14508–14513.

Raiber, E. A., Murat, P., Chirgadze, D. Y., Beraldi, D., Luisi, B. F., & Balasubramanian, S. (2015). 5-Formylcytosine alters the structure of the DNA double helix. *Nature Structural & Molecular Biology, 22*(1), 44–49. http://dx.doi.org/10.1038/nsmb.2936.

Rajasethupathy, P., Antonov, I., Sheridan, R., Frey, S., Sander, C., Tuschl, T., et al. (2012). A role for neuronal piRNAs in the epigenetic control of memory-related synaptic plasticity. *Cell, 149*(3), 693–707. http://dx.doi.org/10.1016/j.cell.2012.02.057.

Rajasethupathy, P., Sankaran, S., Marshel, J. H., Kim, C. K., Ferenczi, E., Lee, S. Y., et al. (2015). Projections from neocortex mediate top-down control of memory retrieval. *Nature, 526*(7575), 653–659. http://dx.doi.org/10.1038/nature15389.

Razin, A., & Friedman, J. (1981). DNA methylation and its possible biological roles. *Progress in Nucleic Acid Research and Molecular Biology, 25*, 33–52.

Reik, W., Dean, W., & Walter, J. (2001). Epigenetic reprogramming in mammalian development. *Science, 293*(5532), 1089–1093. http://dx.doi.org/10.1126/science.1063443.

Roberson, E. D., & Sweatt, J. D. (2001). Memory-forming chemical reactions. *Reviews in the Neurosciences, 12*(1), 41–50.

Robertson, K. D., Ait-Si-Ali, S., Yokochi, T., Wade, P. A., Jones, P. L., & Wolffe, A. P. (2000). DNMT1 forms a complex with Rb, E2F1 and HDAC1 and represses transcription from E2F-responsive promoters. *Nature Genetics, 25*(3), 338–342. http://dx.doi.org/10.1038/77124.

Roth, E. D., Roth, T. L., Money, K. M., SenGupta, S., Eason, D. E., & Sweatt, J. D. (2015). DNA methylation regulates neurophysiological spatial representation in memory formation. *Neuroepigenetics, 2*, 1–8. http://dx.doi.org/10.1016/j.nepig.2015.03.001.

Rudenko, A., Dawlaty, M. M., Seo, J., Cheng, A. W., Meng, J., Le, T., et al. (2013). Tet1 is critical for neuronal activity-regulated gene expression and memory extinction. *Neuron, 79*(6), 1109–1122. http://dx.doi.org/10.1016/j.neuron.2013.08.003.

Schultz, W., Dayan, P., & Montague, P. R. (1997). A neural substrate of prediction and reward. *Science, 275*(5306), 1593–1599.

Song, C. X., Szulwach, K. E., Dai, Q., Fu, Y., Mao, S. Q., Lin, L., et al. (2013). Genome-wide profiling of 5-formylcytosine reveals its roles in epigenetic priming. *Cell, 153*(3), 678–691. http://dx.doi.org/10.1016/j.cell.2013.04.001.

Stuber, G. D., Klanker, M., de Ridder, B., Bowers, M. S., Joosten, R. N., Feenstra, M. G., et al. (2008). Reward-predictive cues enhance excitatory synaptic strength onto midbrain dopamine neurons. *Science, 321*(5896), 1690–1692.

Sultan, F. A., & Sweatt, J. D. (2013). The role of the gadd45 family in the nervous system: a focus on neurodevelopment, neuronal injury, and cognitive neuroepigenetics. *Advances in Experimental Medicine and Biology, 793*, 81–119. http://dx.doi.org/10.1007/978-1-4614-8289-5_6.

Sultan, F. A., Wang, J., Tront, J., Liebermann, D. A., & Sweatt, J. D. (2012). Genetic deletion of Gadd45b, a regulator of active DNA demethylation, enhances long-term memory and synaptic plasticity. *The Journal of Neuroscience: The Official Journal of the Society for Neuroscience, 32*(48), 17059–17066. http://dx.doi.org/10.1523/JNEUROSCI.1747-12.2012.

Sutcliffe, J. S., Nelson, D. L., Zhang, F., Pieretti, M., Caskey, C. T., Saxe, D., et al. (1992). DNA methylation represses FMR-1 transcription in fragile X syndrome. *Human Molecular Genetics, 1*(6), 397–400.

Tahiliani, M., Koh, K. P., Shen, Y., Pastor, W. A., Bandukwala, H., Brudno, Y., et al. (2009). Conversion of 5-methylcytosine to 5-hydroxymethylcytosine in mammalian DNA by MLL partner TET1. *Science, 324*(5929), 930–935.

Tsai, Y. P., Chen, H. F., Chen, S. Y., Cheng, W. C., Wang, H. W., Shen, Z. J., et al. (2014). TET1 regulates hypoxia-induced epithelial-mesenchymal transition by acting as a co-activator. *Genome Biology, 15*(12), 513. http://dx.doi.org/10.1186/s13059-014-0513-0.

Tsai, H. C., Zhang, F., Adamantidis, A., Stuber, G. D., Bonci, A., de Lecea, L., et al. (2009). Phasic firing in dopaminergic neurons is sufficient for behavioral conditioning. *Science, 324*(5930), 1080–1084.

Tse, D., Takeuchi, T., Kakeyama, M., Kajii, Y., Okuno, H., Tohyama, C., et al. (2011). Schema-dependent gene activation and memory encoding in neocortex. *Science, 333*(6044), 891–895. http://dx.doi.org/10.1126/science.1205274.

Turrigiano, G. G., Leslie, K. R., Desai, N. S., Rutherford, L. C., & Nelson, S. B. (1998). Activity-dependent scaling of quantal amplitude in neocortical neurons. *Nature, 391*(6670), 892–896. http://dx.doi.org/10.1038/36103.

Vanyushin, B. F., Tushmalova, N. A., & Guskova, L. V. (1974). Brain DNA methylation as an indicator of genome participation in the individually acquired memory mechanisms. *Doklady Akademii Nauk USSR, 219*(3), 742–744.

Vanyushin, B. F., Tushmalova, N. A., & Guskova, L. V. (1977). Changes in rat brain DNA methylation following conditional memory formation. *Molecular Biology (Moscow), 11*, 181–188.

Wu, S. C., & Zhang, Y. (2010). Active DNA demethylation: many roads lead to Rome. *Nature Reviews. Molecular Cell Biology, 11*(9), 607–620.

Xu, Y., Xu, C., Kato, A., Tempel, W., Abreu, J. G., Bian, C., et al. (2012). Tet3 CXXC domain and dioxygenase activity cooperatively regulate key genes for Xenopus eye and neural development. *Cell, 151*(6), 1200–1213. http://dx.doi.org/10.1016/j.cell.2012.11.014.

Yu, H., Su, Y., Shin, J., Zhong, C., Guo, J. U., Weng, Y. L., et al. (2015). Tet3 regulates synaptic transmission and homeostatic plasticity via DNA oxidation and repair. *Nature Neuroscience, 18*(6), 836–843. http://dx.doi.org/10.1038/nn.4008.

Zhang, R. R., Cui, Q. Y., Murai, K., Lim, Y. C., Smith, Z. D., Jin, S., et al. (2013). Tet1 regulates adult hippocampal neurogenesis and cognition. *Cell Stem Cell, 13*(2), 237–245. http://dx.doi.org/10.1016/j.stem.2013.05.006.

CHAPTER 7

Measuring CpG Methylation by SMRT Sequencing

Y. Suzuki[1], J. Korlach[2], S. Morishita[1]
[1]The University of Tokyo, Tokyo, Japan; [2]Pacific Biosciences, Menlo Park, CA, United States

DNA METHYLOME OF DISEASE-ASSOCIATED REPEATS

There has been a great deal of interest in identification of genome-wide epigenetic DNA modifications in recent years, because DNA modifications play an essential role in cellular and developmental processes (Anway, Cupp, Uzumcu, & Skinner, 2005; Jirtle & Skinner, 2007; Miller, 2010; Molaro et al., 2011; Qu et al., 2012; Schmitz et al., 2011; Smith et al., 2012; Weaver et al., 2004; Zemach, McDaniel, Silva, & Zilberman, 2010). Some human transposable elements (TEs), such as long interspersed nuclear elements (LINEs), are reported to transpose actively within somatic cells along differentiation of neural tissues, and to be partly regulated by DNA methylation (Muotri et al., 2005, 2010). One study showed that each family of human TEs is in a variety of methylation statuses according to tissue type by looking at the mixture of methylation information on the consensus sequence of TEs in the same family (Xie et al., 2013). Many human diseases are associated with the disruption of DNA modifications. In particular, hypomethylation of repetitive elements, such as LINE-1 elements, has been also related to some cancers (Ross, Rand, & Molloy, 2010; Wilson, Power, & Molloy, 2007). Although only a few LINE-1 elements exhibit activity in the human genome (Beck et al., 2010), transpositions of these elements have been reported in various cancer genomes (Goodier, 2014; Lee et al., 2012), and importantly, it has been reported that transpositions are correlated with hypomethylation of the promoter region of LINE-1 elements (Tubio et al., 2014). Therefore, it is essential to develop an experimental framework that can characterize the methylation state of repetitive elements in a genome-wide manner.

TRADITIONAL METHODS FOR OBSERVING DNA METHYLOME

Bisulfite treatment converts cytosine to uracil while leaving 5-methylcytosine (5mC) unchanged (Hayatsu, Wataya, Kai, & Iida, 1970; Shapiro, Servis, & Welcher, 1970). After bisulfite conversion, sequencing PCR-amplified DNA fragments allows us to observe the methylation states of individual cytosines at a single-base resolution because unmethylated cytosines are converted to thymines in the sequences (Frommer et al., 1992). This

DNA Modifications in the Brain
ISBN 978-0-12-801596-4
http://dx.doi.org/10.1016/B978-0-12-801596-4.00007-1

technique is now called bisulfite sequencing, and a variety of improved methods were proposed before the development of genome-wide detection methods (Fraga and Esteller, 2002).

For detecting genome-wide CpG methylation, methylated DNA immune-precipitation (MeDIP) has been developed, and its key idea is to isolate methylated DNA fragments by using 5mC antibodies (Weber et al., 2005). Microarray-based MeDIP methods have been widely used for monitoring DNA methylation states of a limited set of selected CpG sites such as CpG islands of RefSeq genes at a reasonable cost (Weber et al., 2005; Zhang et al., 2006); however, they do not provide a comprehensive view of CpG sites. Moreover, they are not designed to observe CpG sites in highly repetitive regions because of the difficulty in interrogating these CpG sites.

The advent of second-generation sequencing technology has increased the efficiency of the generation of precise genome-wide methylation maps at a single-base resolution by using bisulfite treatment (Cokus et al., 2008; Harris et al., 2010; Lister et al., 2008, 2009; Meissner et al., 2008; Miura, Enomoto, Dairiki, & Ito, 2012) or by using MeDIP-sequencing (Down et al., 2008); however, these sequencing-based technologies have difficulty in characterizing the methylation status of CpGs in regions that are highly similar to other regions. Bisulfite-treated short reads from these regions often fail to map uniquely to their original positions; instead, they are likely to be aligned ambiguously with multiple positions. Moreover, first- and second-generation sequencing technology often fails to sequence DNA regions with a GC content >60% (Aird et al., 2011) and may exhibit bias against GC-rich regions. These inherent problems of second-generation sequencing may result in underrepresentation of methylation information on specific DNA regions, such as TEs and low-complexity repeat sequences (Bock et al., 2010; Gifford et al., 2013; Harris et al., 2010; Jiang et al., 2013; Lister et al., 2009). Especially, the younger and more active transposons are thought to retain higher fidelity and are therefore difficult to address using short reads.

SMRT SEQUENCING TO DETECT DNA MODIFICATIONS

DNA polymerase is used to perform single-molecule real-time (SMRT) sequencing (Eid et al., 2009; Korlach et al., 2008), and this system is capable of sequencing reads of an average length of ~10 kb. SMRT sequencing is also able to sequence genomic regions with extremely high GC contents. A striking example is a previous report of the sequencing of a >2-kb region with a GC content of 100% (Loomis et al., 2012), indicating that SMRT sequencing is less vulnerable to sequence composition bias than is first- and second-generation sequencing. SMRT sequencing of bisulfite-treated DNA fragments may allow identification of DNA methylation within long regions; however, this approach is not promising because bisulfite treatment divides DNA into short fragments <1000 bp (Miura et al., 2012).

Another advantage of SMRT sequencing is the direct detection of DNA modifications. In SMRT sequencing, we can observe the base sequence in a single DNA molecule as each corresponding nucleotide is incorporated using the time course of the fluorescence pulses. From this time course information, we can determine the interpulse duration (IPD), defined as the time interval separating the pulses of two neighboring bases. Importantly, the IPD of the same genomic position varies and has a significant and predictable response to DNA modifications due to the sensitivity of DNA polymerase kinetics to DNA modifications and damage. Consequently, the IPD ratio (IPDR), the ratio of the average IPD in DNA templates with modifications to that in control templates, tends to be perturbed systematically, allowing identification of DNA modifications.

Indeed, SMRT sequencing methods have been used to detect changes in 5-hydroxymethylcytosine (Flusberg et al., 2010), N4-methylcytosine (Clark et al., 2012), and N6-methyladenine (Fang et al., 2012; Feng et al., 2013; Flusberg et al., 2010), as well as damaged DNA bases (Clark, Spittle, Turner, & Korlach, 2011) in bacteria and mitochondria; however, estimation of 5mC residues by using low-coverage reads is prone to errors and requires extensive coverage at each position to clarify the base-wise 5mC state and therefore becomes costly (Fang et al., 2012; Flusberg et al., 2010; Schadt et al., 2012). Clark et al. (2013) attempted to improve the detection of microbial 5mC in the *Escherichia coli* and *Bacillus halodurans* genomes by using Tet1-mediated oxidation to convert 5mC into 5-carboxylcytosine in SMRT reads of ~150× coverage per DNA strand. Kinetic information for low-coverage SMRT reads at a single CpG site is not reliable for predicting the methylation status.

It is notable that unmethylated CpG dinucleotides are rare (~10%) in vertebrates and generally do not exist in isolation, but often range over long hypomethylated regions (Bock, Walter, Paulsen, & Lengauer, 2008; Eckhardt et al., 2006; Gifford et al., 2013; Nautiyal et al., 2010; Qu et al., 2012; Shoemaker, Deng, Wang, & Zhang, 2010; Xie et al., 2013). Su et al. (2012) reported that the average length of unmethylated regions in five human cell types is ~2 kb. Thus, estimating regions of hypomethylated CpG sites is informative in most cases. Similarly, integrating kinetic information for many CpG sites in a long region can increase the confidence in detecting methylation when the status of those sites is correlated and shows promise for predicting the methylation status in a block by using low-coverage SMRT reads. Therefore, we examined the feasibility of the approach and present a novel computational algorithm that integrates SMRT sequencing kinetic data and determines the methylation statuses of CpG sites.

SMRT sequencing is unique in long read output, which has been shown to be useful in a variety of applications such as sequencing of bacterial genomes (Bashir et al., 2012; Zhang et al., 2012), closing gaps in draft genomes (English et al., 2012), de novo assembly of unknown genomes (Chaisson et al., 2014; Koren et al., 2012; Pendleton et al., 2015), sequencing of giant short tandem repeats (eg, CGG repeats) (Loomis et al., 2012), and the comprehensive characterization of mRNA isoforms (Au et al., 2013). Based on these observations and approaches, we examined the possibility of determining the methylation status of highly similar occurrences of TEs in human and medaka fish (*Oryzias latipes*), which could be investigated only using long reads.

PREDICTION OF THE REGIONAL METHYLATION STATE FROM KINETIC DATA

Fig. 7.1A shows a schematic representation of the basic concept of our method. First, as a raw ingredient for prediction, we defined the IPDR profile of a CpG site as an array of IPDR measurements of 21 bp surrounding the CpG site. With low coverage, the IPDR profiles at individual CpG sites are noisy and insufficient for determining whether the focal CpG site is unmethylated or methylated. However, if we could somehow identify the boundaries of hyper- and hypomethylated regions, it would be possible to take the average of the IPDR profile for the CpGs within each region and would allow better prediction of the methylation state of each region from its average IPDR profile, which has less noise than the profile of a single CpG site.

We implemented our method using linear discrimination of the vectors of (average) IPDR profiles around the focal CpG sites. We represented the vectors as points residing in the Euclidean space of the appropriate dimension and attempted to separate the points by a decision hyperplane. For better accuracy, we optimized two parameters of the

Figure 7.1 Schema illustrating our integration method. (A) The top three distributions show the typical interpulse duration ratio (IPDR) profiles within 10 bp of the CpG sites in the raw data. The IPDR profiles of individual CpG sites were treated as points in the 21-dimensional (dim) feature space. Red-colored unmethylated CpGs and blue-colored methylated CpGs are difficult to separate using a hyperplane. Therefore, initially, we had little knowledge about the methylation status of each CpG site from the raw data, as illustrated by the question marks at the CpG sites. Our algorithm predicts the boundary of hypo- and hypermethylated CpG sites. The average IPDR profiles of the two regions, which have clearly distinct IPDR profiles, are shown below the two regions separated by the boundary. Red circles and blue boxes represent unmethylated and methylated CpGs, respectively, predicted by our algorithm (annotated as predicted regions) and were observed by bisulfite sequencing (labeled answer). In the feature space, red and blue disks represent the IPDR profiles of predicted regions. (B) Comparison of our prediction with the available human genome methylome data. From top to bottom, below the RefSeq gene track, black bars indicate hypomethylated regions predicted from single-molecule real-time sequencing data using our method. Yellow and black bars show the methylation level and read coverage obtained from public bisulfite sequencing data, respectively, and blue boxes show hypomethylated regions predicted from the bisulfite data. Green bars below indicate the alignability of short (100-bp) reads. The bottom row shows repeat masker tracks. *LINE*, long interspersed nuclear element; *LTR*, long terminal repeat; *SINE*, short interspersed nuclear element. (C) Accuracy of our method. The sensitivity and precision (proportion of true positives among the predicted positives) are evaluated on individual CpG sites when we change the intercept of the hyperplane and set the minimum number of CpG sites in a region, B, to 30, 35, 40, 45, and 50. (D) Receiver operating characteristic curve of false-positive rate and sensitivity.

(A)

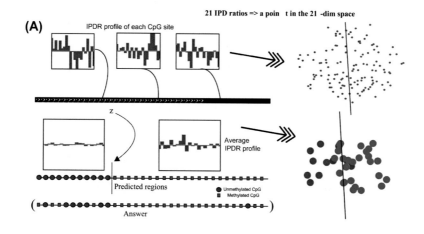

(B)

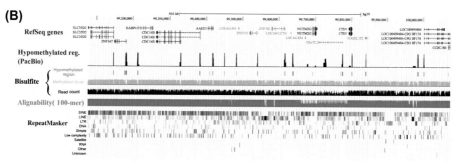

(C)

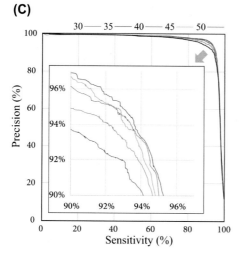

(D)

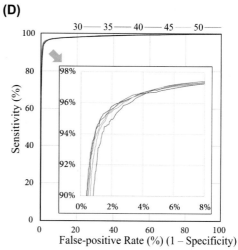

decision hyperplane: beta (orientation) and gamma (intercept). As hypomethylated regions are ~2 kb in size (on average) and contain ~50 CpG sites in vertebrate genomes (Su et al., 2012), in the prediction, we assumed unmethylated regions to have at least b CpG sites and integrated the IPDR profiles to make predictions, which was effective in reducing noise in the IPD measurements. Although setting lower bound b to 50 is supported by the plausible heuristics with a biological basis, a looser bound ($b < 50$) allows us to detect shorter regions. We therefore examined whether we could use a smaller value of b ($= 30, 35, 40, 45$) without degrading the accuracy of prediction. Finally, our method divides the genome into regions containing $\geq b$ CpG sites, such that each region is either hypomethylated or hypermethylated. An example of our prediction for the human genome is shown in Fig. 7.1B, in which gene promoter hypomethylation was captured correctly.

To determine whether our strategy is effective and its dependence on the amount of data available, we performed predictions using five medaka datasets with different read coverage, and we determined the depth of coverage that would be sufficient to correctly detect unmethylated CpG sites. We calculated various accuracy measures, such as sensitivity (recall), specificity ($1 -$ false-positive rate), and precision by comparison our prediction on each CpG site with the methylation level determined in a bisulfite sequencing study (Qu et al., 2012). We made the trade-off between sensitivity and precision through the selection of gamma (the intercept of the decision hyperplane). As most CpG sites in the medaka genome are methylated consistently, there are only a small number of positive examples of unmethylated CpGs, and therefore, precision is more informative than specificity in evaluation. Our prediction achieved 93.7% sensitivity and 93.9% precision with a 29.9-fold mapped read coverage, or 93.0% sensitivity and 94.9% precision depending on the selection of the intercept (Fig. 7.1C). We also compared our predictions made on our human sample to the beta value [an indicator of methylation level expressed as a value ranging over (0,1)] obtained by Illumina BeadChip analysis. We confirmed the strong correlation between our predictions and beta values. We can also extend our method to determine intermediate methylation states ranging from 0 to 1 by changing the value of the intercept.

GENOME-WIDE METHYLATION PATTERN OF REPETITIVE ELEMENTS IN THE HUMAN GENOME

We investigated how individual occurrences of repetitive elements were methylated in the human genome, as summarized in Fig. 7.2A. Fractions of hypomethylated repeat occurrences vary considerably among different classes of repetitive elements, from ~1% for L1 and Alu to ~50% for MIR and >70% for simple repeats and low-complexity regions. To validate our prediction regarding the repeat occurrences, we selected 21 regions for bisulfite Sanger sequencing, designed primers for nested PCR, and could

Class	With >9 CpGs (a)	Covered (b)	b/a	Covered with >5× (c)	c/a	Hypomethylated (d)	d/c
LINE/L1	50,795	50,127	98.7	45,379	89.3	356	0.8
LINE/L2	4977	4961	99.7	4637	93.2	244	5.3
LINE/CR1	178	178	100.0	165	92.7	5	3.0
LINE/RTE-X	65	64	98.5	60	92.3	1	1.7
SINE/Alu	238,701	235,527	98.7	214,341	89.8	2282	1.1
SINE/MIR	374	371	99.2	343	91.7	169	49.3
LTR/ERV1	19,638	19,354	98.6	17,739	90.3	348	2.0
LTR/ERVK	5175	5079	98.1	4603	88.9	87	1.9
LTR/ERVL	4395	4350	99.0	3991	90.8	82	2.1
LTR/ERVL-MaLR	4366	4327	99.1	3933	90.1	69	1.8
LTR/Gypsy	108	104	96.3	89	82.4	9	10.1
Retroposon/SVA	2906	2796	96.2	2427	83.5	3	0.1
DNA/hAT-Blackjack	83	83	100.0	75	90.4	2	2.7
DNA/hAT-Charlie	1460	1452	99.5	1342	91.9	55	4.1
DNA/hAT-Tip100	326	322	98.8	305	93.6	19	6.2
DNA/MULE-MuDR	92	92	100.0	89	96.7	2	2.2
DNA/PiggyBac	57	55	96.5	52	91.2	1	1.9
DNA/TcMar-Mariner	384	384	100.0	360	93.8	1	0.3
DNA/TcMar-Tigger	2821	2801	99.3	2649	93.9	43	1.6
rRNA	68	66	97.1	66	97.1	8	12.1
Simple_repeat	6256	6191	99.0	5434	86.9	3849	70.8
Low_complexity	1068	1064	99.6	942	88.2	789	83.8

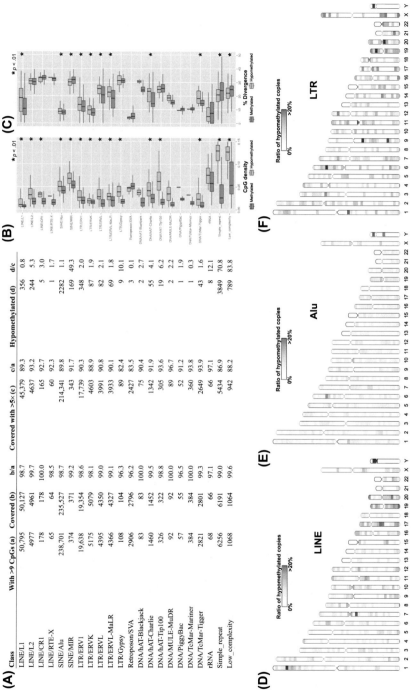

Figure 7.2 Epigenetic landscape of repetitive elements in the human genome. (A) Summary of methylation status on repetitive elements. Because some occurrences of repetitive elements contain no or very few CpG sites, we only consider those occurrences with at least 10 CpGs to exclude other less informative cases (a). First, we checked whether single-molecule real-time (SMRT) reads could address the repetitive regions in a useful manner for methylation analysis. Specifically, we considered a repeat occurrence to be covered by uniquely mapped SMRT reads if the interpulse duration ratio was available on ≥50% of CpGs (b), and found that >96% were covered for every repeat type. To draw robust conclusions, we further applied a stringent quality control process to each repeat occurrence such that the read coverage was >5 (c). Although this step reduced the number of repeat occurrences under consideration by 3–18%, this reduction could be mitigated simply by producing more data. Finally, we treated an occurrence as hypomethylated if ≥50% of CpGs were predicted as hypomethylated (d). (B and C) Distribution of CpG density (B) and sequence divergence from the representative in each repeat class (C) for methylated (cyan) and hypomethylated (pink) repeat occurrences. The asterisks indicate statistical significance (p < 1%) determined by the U test. (D–F) Genome-wide distribution of hypomethylated repetitive elements. The ratio of hypomethylated repeat occurrences to all occurrences in each 5-Mb bin is indicated by color shadings.

amplify six regions, indicating the difficulty in observing DNA methylation of repetitive elements by using traditional bisulfite Sanger sequencing. In the six amplified regions, we confirmed the consistency between our prediction and the methylation state observed by bisulfite Sanger sequencing.

We then examined the features for characterizing the differences between hyper-methylated and hypomethylated repetitive elements. First, CpG density was significantly higher in the hypomethylated occurrences in almost all classes of repetitive elements ($p < 1\%$, Fig. 7.2B). This observation was consistent with the known association between CpG-rich regions and hypomethylation because hypermethylation leads to depletion of CpG sites through deamination (Cooper & Krawczak, 1989). Second, sequence divergence from the representative in each repeat class also showed a correlation with methylation status (Fig. 7.2C). For most classes, with the apparent exception of simple repeats, low-complexity regions, and MIR elements, hypomethylated occurrences were significantly more divergent than were hypermethylated occurrences ($p < 1\%$, Fig. 7.2C), presumably because younger copies of a repeat element are less divergent and are likely to be targets of DNA methylation.

Next, we examined whether the hypomethylated repeat occurrences were distributed uniformly or nonuniformly throughout the entire genome. We selected three major classes of repetitive elements for this analysis: long interspersed nuclear element (LINE), Alu, and long terminal repeat (LTR). We calculated the ratios of hypomethylated copies to all repetitive elements in individual nonoverlapping bins 5 Mb in size (Fig. 7.2D–F). The nonrandom distribution patterns were more evident for LINE and LTR than for Alu. For example, we found hypomethylated LINEs to be enriched in the p-arm of chromosome 1 and in chromosomes 17 and 19. There were hypomethylation "hot spots" of LTR elements, for example, in chromosomes 6 and 9. It is intriguing that some of these hypomethylation hot spots, such as those in the p-arms of chromosomes 6 and Y, seem to be shared among different classes of repetitive elements.

ANALYSIS OF AN ACTIVE TRANSPOSABLE ELEMENT

The medaka has an innate autonomous transposon known as *Tol2*, which is one of the first examples of autonomous transposons in vertebrate genomes and a useful tool for genetic engineering of vertebrates, such as zebrafish and mice (Kawakami, 2007). The excision activities of *Tol2* are promoted when DNA methylation is reduced by 5-azacytidine treatment, which suggests that DNA methylation is one of the mechanisms regulating the *Tol2* transposition (Iida et al., 2006). Nevertheless, observing the methylation status of each *Tol2* copy by using short reads is difficult, because *Tol2* is 4682 bp in length, and ~20 copies of *Tol2* exist in the genome, all of which are essentially identical (>99.8%).

To elucidate the methylation status of each *Tol2* copy, we applied our method to a new assembly of the Hd-rR genome obtained exclusively from SMRT reads. We found

17 copies of *Tol2* contained entirely within this assembly, and then called the methylation status of these *Tol2*. For comparison, we mapped bisulfite-treated short reads to these contigs and determined the methylation level. The methylation status of these *Tol2*, observed by SMRT reads and bisulfite sequencing, are shown in Fig. 7.3. Although virtually no *Tol2* copies were mapped by bisulfite reads, as expected from their extremely high fidelity, 16 out of 17 copies were anchored by SMRT reads, and all were predicted to be hypermethylated by our method. For the regions examined by both SMRT reads and bisulfite-treated short reads, our prediction was consistent with the methylation level calculated from the bisulfite-treated reads. For example, one *Tol2* copy was surrounded by hypomethylated regions (number 14). From the bisulfite data, it seemed that the body of *Tol2*, from which data were missing, was hypomethylated. Nevertheless, our prediction estimated this region to be hypermethylated. These results demonstrate the ability of our method to clarify DNA methylation states of highly identical repetitive elements such as active transposons.

DISCUSSION

We addressed the problem of uncovering the landscape of DNA methylation of repetitive elements. To this end, we described a unique application of SMRT sequencing to epigenetics. This direction had been already explored in the research community for bacterial and viral species. However, this application in large vertebrate genomes has been largely unexplored because of the subtle cytosine methylation signals in the kinetic information.

Our method uses relatively small amounts of kinetic information by incorporating a model reflecting our prior knowledge on the regional patterns of CpG methylation of vertebrate genomes. We confirmed the validity of our strategy by comparing the prediction to bisulfite sequencing data on medaka and to BeadChip analysis on human samples. These two datasets had very different characteristics, which seemed to be partly because of the methods used and partly because of the nature of the samples used (ie, the medaka samples were derived from an inbred strain, whereas the human samples were from diploid cells). Despite such differences in characteristics, our method using the same parameters performed almost equally well for both datasets. These observations suggested that the choice of parameters is robust for a wide variety of samples, which is a desirable feature for any method.

Our method had important strengths compared with conventional tools for epigenetic studies, such as bisulfite sequencing or affinity-based assays, with not only an expected increase in comprehensiveness by virtue of long SMRT reads but also in the remarkable reduction of laboratory work. If an epigenetic study is conducted alongside a resequencing study or a de novo assembly study using SMRT sequencing, the methylation status could be called solely in silico, and no additional experiments would be necessary.

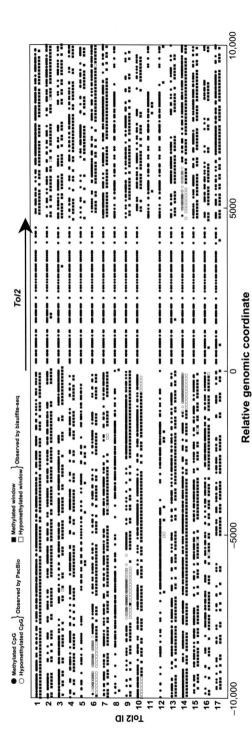

Figure 7.3 Methylation analysis of *Tol2*, a 4682-bp autonomous transposon, in medaka. The new genome assembly of single-molecule real-time (SMRT) reads had 17 regions (contigs) that contained complete *Tol2* copies. Circles show our prediction of the methylation state of CpG sites, whereas rectangles show the methylation states within each 100-bp window obtained from bisulfite sequencing. For both tracks, *red/outlined and blue/filled* indicate hypomethylation and hypermethylation, respectively. As the 11th region was located at the extreme of the contig, *Tol2* was not observed successfully by either SMRT sequencing or bisulfite sequencing. For the other 16 regions, hypermethylation of *Tol2* was observed consistently by SMRT sequencing, whereas virtually no information was available on the *Tol2* region from bisulfite sequencing.

REFERENCES

Aird, D., Ross, M. G., Chen, W.-S., Danielsson, M., Fennell, T., Russ, C., et al. (2011). Analyzing and minimizing PCR amplification bias in Illumina sequencing libraries. *Genome Biology, 12*, R18.

Anway, M. D., Cupp, A. S., Uzumcu, M., & Skinner, M. K. (2005). Epigenetic transgenerational actions of endocrine disruptors and male fertility. *Science, 308*, 1466–1469.

Au, K. F., Sebastiano, V., Afshar, P. T., Durruthy, J. D., Lee, L., Williams, B. A., et al. (2013). Characterization of the human ESC transcriptome by hybrid sequencing. *Proceedings of the National Academy of Sciences of the United States of America, 110*, E4821–E4830.

Bashir, A., Klammer, A., Robins, W. P., Chin, C.-S., Webster, D., Paxinos, E., et al. (2012). A hybrid approach for the automated finishing of bacterial genomes. *Nature Biotechnology, 30*, 701–707.

Beck, C. R., Collier, P., Macfarlane, C., Malig, M., Kidd, J. M., Eichler, E. E., et al. (2010). LINE-1 retrotransposition activity in human genomes. *Cell, 141*, 1159–1170.

Bock, C., Tomazou, E. M., Brinkman, A. B., Müller, F., Simmer, F., Gu, H., et al. (2010). Quantitative comparison of genome-wide DNA methylation mapping technologies. *Nature Biotechnology, 28*, 1106–1114.

Bock, C., Walter, J., Paulsen, M., & Lengauer, T. (2008). Inter-individual variation of DNA methylation and its implications for large-scale epigenome mapping. *Nucleic Acids Research, 36*, e55.

Chaisson, M. J. P., Huddleston, J., Dennis, M. Y., Sudmant, P. H., Malig, M., Hormozdiari, F., et al. (2014). Resolving the complexity of the human genome using single-molecule sequencing. *Nature, 517*, 608–611.

Clark, T. A., Lu, X., Luong, K., Dai, Q., Boitano, M., Turner, S., et al. (2013). Enhanced 5-methylcytosine detection in single-molecule, real-time sequencing via Tet1 oxidation. *BMC Biology, 11*, 4.

Clark, T. A., Murray, I. A., Morgan, R. D., Kislyuk, A. O., Spittle, K. E., Boitano, M., et al. (2012). Characterization of DNA methyltransferase specificities using single-molecule, real-time DNA sequencing. *Nucleic Acids Research, 40*, e29.

Clark, T. A., Spittle, K. E., Turner, S. W., & Korlach, J. (2011). Direct detection and sequencing of damaged DNA bases. *Genome Integrity, 2*, 10.

Cokus, S. J., Feng, S. H., Zhang, X.Y., Chen, Z. G., Merriman, B., Haudenschild, C. D., et al. (2008). Shotgun bisulphite sequencing of the Arabidopsis genome reveals DNA methylation patterning. *Nature, 452*, 215–219.

Cooper, D., & Krawczak, M. (1989). Cytosine methylation and the fate of CpG dinucleotides in vertebrate genomes. *Human Genetics, 83*, 181.

Down, T. A., Rakyan, V. K., Turner, D. J., Flicek, P., Li, H., Kulesha, E., et al. (2008). A Bayesian deconvolution strategy for immunoprecipitation-based DNA methylome analysis. *Nature Biotechnology, 26*(7), 779–785.

Eckhardt, F., Lewin, J., Cortese, R., Rakyan, V. K., Attwood, J., Burger, M., et al. (2006). DNA methylation profiling of human chromosomes 6, 20 and 22. *Nature Genetics, 38*, 1378–1385.

Eid, J., Fehr, A., Gray, J., Luong, K., Lyle, J., Otto, G., et al. (2009). Real-time DNA sequencing from single polymerase molecules. *Science (New York, NY), 323*, 133–138.

English, A. C., Richards, S., Han, Y., Wang, M., Vee, V., Qu, J., et al. (2012). Mind the gap: upgrading genomes with Pacific Biosciences RS long-read sequencing technology. *PLoS One, 7*, e47768.

Fang, G., Munera, D., Friedman, D. I., Mandlik, A., Chao, M. C., Banerjee, O., et al. (2012). Genome-wide mapping of methylated adenine residues in pathogenic *Escherichia coli* using single-molecule real-time sequencing. *Nature Biotechnology, 30*(12).

Feng, Z., Fang, G., Korlach, J., Clark, T., Luong, K., Zhang, X., et al. (2013). Detecting DNA modifications from SMRT sequencing data by modeling sequence context dependence of polymerase kinetic. *PLoS Computational Biology, 9*, e1002935.

Flusberg, B. A., Webster, D. R., Lee, J. H., Travers, K. J., Olivares, E. C., Clark, T. A., et al. (2010). Direct detection of DNA methylation during single-molecule, real-time sequencing. *Nature Methods, 7*, 461–465.

Fraga, M. F., & Esteller, M. (2002). DNA methylation: a profile of methods and applications. *Biotechniques, 33*, 632–649.

Frommer, M., McDonald, L. E., Millar, D. S., Collis, C. M., Watt, F., Grigg, G. W., et al. (1992). A genomic sequencing protocol that yields a positive display of 5-methylcytosine residues in individual DNA strands. *Proceedings of the National Academy of Sciences of the United States of America, 89*(5), 1827–1831.

Gifford, C. A., Ziller, M. J., Gu, H., Trapnell, C., Donaghey, J., Tsankov, A., et al. (2013). Transcriptional and epigenetic dynamics during specification of human embryonic stem cells. *Cell, 153*, 1149–1163.

Goodier, J. L. (2014). Retrotransposition in tumors and brains. *Mobile DNA, 5*, 11.

Harris, R. A., Wang, T., Coarfa, C., Nagarajan, R. P., Hong, C., Downey, S. L., et al. (2010). Comparison of sequencing-based methods to profile DNA methylation and identification of monoallelic epigenetic modifications. In *Nature biotechnology* (Vol. 28) (pp. 1097–1105).

Hayatsu, H., Wataya, Y., Kai, K., & Iida, S. (1970). Reaction of sodium bisulfite with uracil, cytosine, and their derivatives. *Biochemistry, 9*(14), 2858–2865.

Iida, A., Shimada, A., Shima, A., Takamatsu, N., Hori, H., Takeuchi, K., et al. (2006). Targeted reduction of the DNA methylation level with 5-azacytidine promotes excision of the medaka fish Tol2 transposable element. *Genetical Research, 87*, 187–193.

Jiang, L., Zhang, J., Wang, J.-J., Wang, L., Zhang, L., Li, G., et al. (2013). Sperm, but not oocyte, DNA methylome is inherited by zebrafish early embryos. *Cell, 153*, 773–784.

Jirtle, R. L., & Skinner, M. K. (2007). Environmental epigenomics and disease susceptibility. *Nature Reviews Genetics, 8*, 253–262.

Kawakami, K. (2007). Tol2: a versatile gene transfer vector in vertebrates. *Genome Biology, 8*, 1–10.

Koren, S., Schatz, M. C., Walenz, B. P., Martin, J., Howard, J. T., Ganapathy, G., et al. (2012). Hybrid error correction and de novo assembly of single-molecule sequencing reads. *Nature Biotechnology, 30*, 693–700.

Korlach, J., Marks, P. J., Cicero, R. L., Gray, J. J., Murphy, D. L., Roitman, D. B., et al. (2008). Selective aluminum passivation for targeted immobilization of single DNA polymerase molecules in zero-mode waveguide nanostructures. *Proceedings of the National Academy of Sciences of the United States of America, 105*, 1176–1181.

Lee, E., Iskow, R., Yang, L., Gokcumen, O., Haseley, P., Luquette, L. J., et al. (2012). Landscape of somatic retrotransposition in human cancers. *Science, 337*, 967–971.

Lister, R., O'Malley, R. C., Tonti-Filippini, J., Gregory, B. D., Berry, C. C., Millar, A. H., et al. (2008). Highly integrated single-base resolution maps of the epigenome in Arabidopsis. *Cell, 133*, 523–536.

Lister, R., Pelizzola, M., Dowen, R. H., Hawkins, R. D., Hon, G., Tonti-Filippini, J., et al. (2009). Human DNA methylomes at base resolution show widespread epigenomic differences. *Nature, 462*, 315–322.

Loomis, E. W., Eid, J. S., Peluso, P., Yin, J., Hickey, L., Rank, D., et al. (2012). Sequencing the unsequenceable: expanded CGG-repeat alleles of the fragile X gene. *Genome Research, 23*(1).

Meissner, A., Mikkelsen, T. S., Gu, H., Wernig, M., Hanna, J., Sivachenko, A., et al. (2008). Genome-scale DNA methylation maps of pluripotent and differentiated cells. *Nature, 454*, 766–770.

Miller, G. (2010). Epigenetics. The seductive allure of behavioral epigenetics. *Science, 329*, 24–27.

Miura, F., Enomoto, Y., Dairiki, R., & Ito, T. (2012). Amplification-free whole-genome bisulfite sequencing by post-bisulfite adaptor tagging. *Nucleic Acids Research, 40*, e136.

Molaro, A., Hodges, E., Fang, F., Song, Q., McCombie, W. R., Hannon, G. J., et al. (2011). Sperm methylation profiles reveal features of epigenetic inheritance and evolution in primates. *Cell, 146*, 1029–1041.

Muotri, A. R., Chu, V. T., Marchetto, M. C., Deng, W., Moran, J. V., & Gage, F. H. (2005). Somatic mosaicism in neuronal precursor cells mediated by L1 retrotransposition. *Nature, 435*, 903–910.

Muotri, A. R., Marchetto, M. C., Coufal, N. G., Oefner, R., Yeo, G., Nakashima, K., et al. (2010). L1 retrotransposition in neurons is modulated by MeCP2. *Nature, 468*, 443–446.

Nautiyal, S., Carlton, V. E., Lu, Y., Ireland, J. S., Flaucher, D., Moorhead, M., et al. (2010). High-throughput method for analyzing methylation of CpGs in targeted genomic regions. *Proceedings of the National Academy of Sciences of the United States of America, 107*, 12587–12592.

Pendleton, M., Sebra, R., Pang, A. W. C., Ummat, A., Franzen, O., Rausch, T., et al. (2015). Assembly and diploid architecture of an individual human genome via single-molecule technologies. *Nature Methods, 12*(8), 780–786.

Qu, W., Hashimoto, S., Shimada, A., Nakatani, Y., Ichikawa, K., Saito, T. L., et al. (2012). Genome-wide genetic variations are highly correlated with proximal DNA methylation patterns. *Genome Research, 22*, 1419–1425.

Ross, J. P., Rand, K. N., & Molloy, P. L. (2010). Hypomethylation of repeated DNA sequences in cancer. *Epigenomics, 2*, 245–269.

Schadt, E. E., Banerjee, O., Fang, G., Feng, Z., Wong, W. H., Zhang, X., et al. (2012). Modeling kinetic rate variation in third generation DNA sequencing data to detect putative modifications to DNA bases. *Genome Research, 23*(1).

Schmitz, R. J., Schultz, M. D., Lewsey, M. G., O'Malley, R. C., Urich, M. A., Libiger, O., et al. (2011). Transgenerational epigenetic instability is a source of novel methylation variants. *Science (New York, NY), 334,* 369–373.

Shapiro, R., Servis, R. E., & Welcher, M. (1970). Reactions of uracil and cytosine derivatives with sodium bisulfite. *Journal of the American Chemical Society, 92*(2), 422–424.

Shoemaker, R., Deng, J., Wang, W., & Zhang, K. (2010). Allele-specific methylation is prevalent and is contributed by CpG-SNPs in the human genome. *Genome Research, 20,* 883–889.

Smith, Z. D., Chan, M. M., Mikkelsen, T. S., Gu, H., Gnirke, A., Regev, A., et al. (2012). A unique regulatory phase of DNA methylation in the early mammalian embryo. *Nature, 484,* 339–344.

Su, J., Yan, H., Wei, Y., Liu, H., Wang, F., Lv, J., et al. (2012). CpG_MPs: identification of CpG methylation patterns of genomic regions from high-throughput bisulfite sequencing data. *Nucleic Acids Research, 41*(1).

Tubio, J. M. C., Li, Y., Ju, Y. S., Martincorena, I., Cooke, S. L., Tojo, M., et al. (2014). Extensive transduction of nonrepetitive DNA mediated by L1 retrotransposition in cancer genomes. *Science, 345.*

Weaver, I. C., Cervoni, N., Champagne, F. A., D'Alessio, A. C., Sharma, S., Seckl, J. R., et al. (2004). Epigenetic programming by maternal behavior. *Nature Neuroscience, 7,* 847–854.

Weber, M., Davies, J. J., Wittig, D., et al. (2005). Chromosome-wide and promoter-specific analyses identify sites of differential DNA methylation in normal and transformed human cells. *Nature Genetics, 37*(8), 853–862.

Wilson, A. S., Power, B. E., & Molloy, P. L. (2007). DNA hypomethylation and human diseases. *Biochimica et Biophysica Acta, 1775,* 138–162.

Xie, M., Hong, C., Zhang, B., Lowdon, R. F., Xing, X., Li, D., et al. (2013). DNA hypomethylation within specific transposable element families associates with tissue-specific enhancer landscape. *Nature Genetics, 45,* 836–841.

Zemach, A., McDaniel, I. E., Silva, P., & Zilberman, D. (2010). Genome-wide evolutionary analysis of eukaryotic DNA methylation. *Science, 328,* 916–919.

Zhang, X., Yazaki, J., Sundaresan, A., Cokus, S., Chan, S. W. -L., Chen, H., et al. (2006). Genome-wide high-resolution mapping and functional analysis of DNA methylation in arabidopsis. *Cell, 126*(6), 1189–1201.

Zhang, X., Davenport, K. W., Gu, W., Daligault, H. E., Munk, A. C., Tashima, H., et al. (2012). Improving genome assemblies by sequencing PCR products with PacBio. *Biotechniques, 53,* 61–62.

CHAPTER 8

Epigenetic Modifications of DNA and Drug Addiction

J. Feng, E.J. Nestler
Icahn School of Medicine at Mount Sinai, New York, NY, United States

INTRODUCTION

DNA methylation is an epigenetic mechanism in which a methyl group is covalently coupled to the C-5 position of cytosine, predominantly at CpG dinucleotides (Jaenisch & Bird, 2003). It is catalyzed by a group of enzymes called DNA methyltransferases (DNMTs), which include the maintenance enzyme DNMT1 and the de novo DNA methyltransferases DNMT3a and DNMT3b. The role of DNA methylation has been widely demonstrated by its involvement in genomic imprinting (silencing of germline specific genes in somatic cells), retroviral silencing, and X chromosome inactivation. These phenomena arise early in development and affect all tissues and cell types.

The role of DNA methylation in the nervous system has unique features because the brain is composed predominantly of postmitotic neurons and slowly dividing glia cells. Accumulating evidence has implicated DNA methylation in neural plasticity, learning and memory, and cognition (Day & Sweatt, 2011; Lubin, Gupta, Parrish, Grissom, & Davis, 2011; Mikaelsson & Miller, 2011; Moore, Le, & Fan, 2013; Nelson & Monteggia, 2011; Shin, Ming, & Song, 2014). Being an aberrant form of neural plasticity (Hyman, Malenka, & Nestler, 2006), drug addiction also has been shown to be mediated, in part, via several forms of epigenetic regulation, such as histone acetylation and methylation, and by noncoding RNA (Feng & Nestler, 2013; Godino, Jayanthi, & Cadet, 2015; Kenny, 2014; Kyzar & Pandey, 2015; LaPlant & Nestler, 2011; Maze & Nestler, 2011; Nielsen, Utrankar, Reyes, Simons, & Kosten, 2012; Ponomarev, 2013; Robison & Nestler, 2011; Rogge & Wood, 2013; Schmidt, McGinty, West, & Sadri-Vakili, 2013; Starkman, Sakharkar, & Pandey, 2012; Walker, Cates, Heller, & Nestler, 2015). However, studies of DNA methylation in addiction are still relatively few in number. Here, we review the literature on DNA methylation and novel forms of DNA modification in addiction, with a focus on cocaine and ethanol, the two drugs of abuse most studied at this level of analysis to date.

DNA Modifications in the Brain
ISBN 978-0-12-801596-4
http://dx.doi.org/10.1016/B978-0-12-801596-4.00008-3

ADDICTION AND REWARD PATHWAY

Addiction can be defined as the loss of control over drug use, and the compulsive seeking and taking of a drug despite adverse effects (Nestler, 2001). It is a serious major medical and social problem worldwide. For example, more than 20 million people in the United States are classified as having a substance use disorder, and among them >1 million are addicted to cocaine (Nielsen, Utrankar, et al., 2012). Approximately 7% or 17 million adults in the United States ages 18 years and older have an alcohol use disorder. In addition, >850,000 adolescents aged 12–17 years have such a diagnosis. Of note, drug taking does not always result in addiction. For example, only ~20% of cocaine users eventually become addicted, and the rate is much lower for alcohol (Anthony, Warner, & Kessler, 1994). As a result there has been a great deal of interest in understanding the transitions from drug use to addiction. Genetic factors are important, with ~50% of the risk for addiction to any drug being genetic. However, the specific genes that comprise this risk remain largely unknown. The other 50% is thought to reflect a range of environmental exposures. Presumably, environmental experience influences addiction risk through epigenetic mechanisms in the brain.

Great progress has been made over the past several decades in identifying brain regions that are important for addiction. The circuit that has received the most attention is referred to as the mesolimbic dopamine system (Kalivas & Volkow, 2011; Nestler, 2001). This system is composed of dopamine neurons in the ventral tegmental area (VTA) of the midbrain projecting to medium spiny neurons in the nucleus accumbens (NAc), a part of ventral striatum. This VTA–NAc circuit is crucial for the recognition of rewards in the environment and for initiating their consumption, but it also responds to aversive stimuli. VTA dopamine neurons innervate many other forebrain regions as well, including hippocampus, amygdala, and prefrontal cortex (PFC), among others. In turn, these cortical and subcortical regions provide glutamatergic innervation of the NAc. Together, these various interconnected circuits are referred to as the brain's reward pathway, crucial for mediating responses to natural rewards, but also the sites where drugs of abuse produce long-lasting changes to underlie addiction (Fig. 8.1).

Rodent models successfully recapitulate key features of drug addiction syndromes seen in humans. The best model is where animals can volitionally self-administer a drug to themselves; a subset of animals, depending on the experimental conditions, become compulsive drug users and show high rates of relapse during abstinence. However, drug self-administration paradigms are very labor intensive, and particularly difficult in mice due to the small caliber of their jugular veins, which hinders intravenous drug access. For these reasons, most studies continue to use experimenter-administered drug [eg, repeated intraperitoneal (IP) injections of cocaine]. Although such passive drug administration paradigms cannot capture

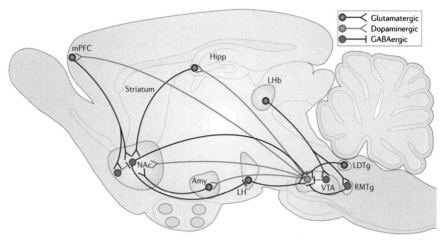

Figure 8.1 *Reward circuitry of the rodent brain.* A simplified schematic of the major dopaminergic, glutamatergic, and GABAergic connections to and from the ventral tegmental area (VTA) and nucleus accumbens (NAc) (*green*, dopaminergic; *red*, glutamatergic; and *blue*, GABAergic) in the rodent brain. The primary reward circuit includes dopaminergic projections from the VTA to the NAc, which release dopamine in response to reward- (and in some cases aversion-) related stimuli. There are also GABAergic projections from the NAc to the VTA; projections through the direct pathway [mediated primarily by D1-type medium spiny neurons (MSNs)] directly innervate the VTA, whereas projections through the indirect pathway (mediated primarily by D2-type MSNs) innervate the VTA via intervening GABAergic neurons in ventral pallidum (not shown). The NAc also contains numerous types of interneurons (not shown). The NAc receives dense innervation from glutamatergic monosynaptic circuits from the medial prefrontal cortex (mPFC) and other prefrontal cortex regions (not shown), hippocampus (Hipp), and amygdala (Amy), among other regions. The VTA receives such inputs from Amy, lateral dorsal tegmentum (LDTg), lateral habenula (LHb), and lateral hypothalamus (LH), among others. These various glutamatergic inputs control aspects of reward-related perception and memory. The glutamatergic circuit from LH to VTA is also mediated by orexin (not shown). RMTg, rostromedial tegmentum (Russo & Nestler, 2013).

consequences of volitional control over drug use, they are rewarding and produce sensitized drug responses. Ultimately, findings from passive drug paradigms must be validated in self-administration models.

DNA METHYLATION AND ITS CATALYZING DNMT ENZYMES IN ADDICTION

Studies of cocaine and other psychostimulants

As noted in the previous section, there remain relatively few studies of DNA methylation in addiction, with most studies to date focused on the role of DNMTs or the methyl-CpG-binding protein 2 (MeCP2). A study from our group demonstrated that repeated cocaine exposure regulated *Dnmt3a* transcription in mouse NAc (LaPlant et al., 2010). Dnmt3a, but not other *Dnmts*, was upregulated at an early time point of withdrawal (4 h after the last cocaine dose), followed by downregulation after 24 h. Whether this surprisingly complicated pattern of *Dnmt3a* regulation is associated with fluctuations in DNA methylation requires further investigation. Importantly; however, after 28 days of withdrawal, after either cocaine IP injections or cocaine self-administration, *Dnmt3a* was upregulated in the NAc (LaPlant et al., 2010). This long-lasting induction of *Dnmt3a* is of particular interest given its potential influence on downstream regulation of target genes, a possibility that requires direct examination. In contrast, different effects of cocaine were reported by Anier, Malinovskaja, Aonurm-Helm, Zharkovsky, and Kalda (2010), who found induction of both *Dnmt3a* and *Dnmt3b* in mouse NAc, but only after acute (not chronic) cocaine administration. The reasons for these discrepancies are not known, but they could be due to the different experimental paradigms used. Furthermore, it has been shown that drug regulation of *Dnmt1* expression in NAc and other regions was dependent on genetic background: chronic methamphetamine treatment increased *Dnmt1* in these regions of Fischer 344/N rats, but exerted the opposite effect in Lewis/N rats (Numachi et al., 2007). This opposite regulation of *Dnmt1* expression was associated with contrasting behavioral susceptibilities to methamphetamine in the two rat lines.

Pharmacological and viral-mediated gene transfer approaches have been used to examine the behavioral influence of DNMTs on drug addiction. Overexpression of DNMT3a in the NAc attenuated the rewarding effects of cocaine (LaPlant et al., 2010). Moreover, DNMT3a overexpression was sufficient to increase dendritic spine density of NAc neurons to comparable levels seen in response to cocaine administration. Conversely, viral-mediated knockdown of DNMT3a in the NAc, or inhibition of DNMTs via local infusion of the DNMT inhibitor RG108, had the opposite effect. These findings establish an important role of DNMT3a in cocaine-induced neural and behavioral plasticity.

MeCP2 is an X-linked methyl-DNA binding protein that is best studied for its involvement in Rett syndrome, an autism spectrum disorder. More recently, MeCP2 has been implicated in neural and behavioral responses to psychostimulants. In one study, MeCP2 was induced in the dorsal striatum of rats with extended access to IV cocaine self-administration. Viral-mediated MeCP2 knockdown in this region decreased the rats' cocaine intake (Im, Hollander, Bali, & Kenny, 2010). In parallel, chronic amphetamine administration was shown to increase MeCP2 phosphorylation at Ser421 in the mouse NAc, and viral-mediated knockdown of MeCP2 in this region increased amphetamine place conditioning, whereas local MeCP2 overexpression had the opposite effect (Deng et al., 2010). Importantly, Deng et al. (2014) subsequently showed that mice with a Ser421Ala mutation in MeCP2 displayed greater locomotor sensitization to experimenter-administered cocaine as well as greater self-administration of the drug. The mutant MeCP2 mice also displayed reduced electrical excitability of NAc medium spiny neurons and altered transcriptional responses to cocaine. These exciting studies together link MeCP2 function in NAc and dorsal striatum with psychostimulant addiction.

The next step in these investigations is to study the effect of drug exposure on DNA methylation itself. Mass spectrometry–based measurements showed that chronic cocaine decreased total levels of methylated DNA in the PFC (Tian et al., 2012). However, there was no such change in the NAc (Feng et al., 2015) or in response to other drugs of abuse, for example, morphine (Tian et al., 2012). A major need in the field is to obtain genome-wide maps of DNA methylation in the NAc and other brain reward regions after chronic drug administration (see later). In the absence of such genome-wide studies, a few candidate genes have been shown to exhibit altered methylation in addiction models (Table 8.1). For example, chronic cocaine administration induced DNA hypermethylation and increased binding of MeCP2 at the protein phosphatase-1 catalytic subunit (*Pp1c*) gene promoter in the NAc, which was associated with transcriptional repression (Anier et al., 2010). In contrast, cocaine administration induced hypomethylation and decreased binding of MeCP2 at the *FosB* promoter in the NAc, associated with induction of FosB (Anier et al., 2010). Wright et al. (2015) showed that chronic cocaine was shown to induce c-Fos expression in the NAc, which was associated with reduced methylation at CpG dinucleotides in the *c-Fos* gene promoter. Outside of brain reward pathways, significant hypomethylation at multiple CpG sites of the *Sox10* promoter region was observed in the corpus callosum of rats at 30 days of forced abstinence from cocaine self-administration (Nielsen, Huang, et al., 2012). As Sox10 expression is enriched in oligodendrocytes, this finding highlights the need to study cocaine regulation of DNA methylation in nonneuronal cell types.

The influence of DNA methylation in addiction models raises the possibility that methylation manipulations might provide a plausible path for addiction therapy. Indeed, methyl supplementation through administration of the methyl donor methionine significantly inhibited cocaine reward in mice (LaPlant & Nestler, 2011; Tian et al., 2012). Also, rats receiving methionine underwent either a sensitization regimen of intermittent cocaine injections or intravenous cocaine self-administration, followed by cue-induced and drug-primed reinstatement. Methionine not only blocked locomotor sensitization but also attenuated drug-primed reinstatement (Wright et al., 2015). Systemic methionine administration was also associated with reversal of DNA hypomethylation at the c-Fos gene in the NAc (Wright et al., 2015). In contrast, using a similar cocaine self-administration approach, intra–NAc injection of a methyl donor promoted cue-induced cocaine seeking after prolonged withdrawal, whereas injection of the DNMT inhibitor RG108 had the opposite effect (Massart et al., 2015). These seemingly opposite effects of a methyl donor on drug behavior may be due to differences in route of administration (systemic vs intra-NAc). Further research is needed to investigate this and alternative possibilities and to test the effect of methyl supplementation in human addicts.

Table 8.1 Examples of candidate genes exhibiting altered DNA methylation in addiction

Gene name	Drug of abuse	Differential methylation region	Direction of change	Associated mRNA/protein change	Species/tissue or cell type	DNA methylation methodology	References
ALDH1A2	Alcohol	Promoter	Hypermethylation	NA	Human/saliva	Illumina bead chip/pyrosequencing	Harlaar et al. (2014)
ANP	Alcohol	Promoter	Hypomethylation	mRNA increase	Human/blood	Methylation specific qPCR	Hillemacher, Frieling, Luber, et al. (2009)
AVP	Alcohol	Promoter	Hypermethylation	mRNA no change	Human/blood	Methylation-specific qPCR	Hillemacher, Frieling, Luber, et al. (2009)
c-Fos	Cocaine	Promoter	Hypomethylation	mRNA increase	Rat/nucleus accumbens	Bisulfite sequencing	Wright et al. (2015)
DAT	Alcohol	Promoter	Hypermethylation	NA	Human/blood	Methylation-specific qPCR	Hillemacher, Frieling, Hartl, et al. (2009)
DLK1	Alcohol	Intergenic region	Hypomethylation	NA	Human/sperm	Bisulfite sequencing	Ouko et al. (2009)
FosB	Cocaine	Promoter	Hypomethylation	mRNA increase	Mouse/nucleus accumbens	MeDIP/methylation-specific qPCR	Anier et al. (2010)
GluA1	Methamphetamine	Promoter	Hypomethylation	Both decrease	Rat/striatum	MeDIP	Jayanthi et al. (2014)
GluA2	Methamphetamine	Promoter	Hypomethylation	Both decrease	Rat/striatum	MeDIP	Jayanthi et al. (2014)
HERP	Alcohol	Promoter	Hypermethylation	mRNA decrease	Human/blood	Methylation-specific qPCR	Bleich et al. (2006)
HTR3A	Alcohol	Promoter	Hypermethylation	NA	Human/blood	Methylation array, Sequenom	H. Zhang et al. (2013)

Gene	Substance	Location	Methylation	Expression	Species/tissue	Method	Reference
MAOA	Alcohol	Promoter	Hypermethylation	Undetectable mRNA	Woman/lymphoblast lines	Sequenom	Philibert, Gunter, Beach, Brody, and Madan (2008)
NGF	Alcohol	Promoter	Hypermethylation	Protein decrease	Human/blood	Bisulfite sequencing	Heberlein et al. (2013)
NR2B	Alcohol	Promoter	Hypomethylation	mRNA increase	Human/blood; Mouse/neuronal culture	Bisulfite sequencing	Biermann et al. (2009) and Marutha Ravindran and Ticku (2005)
OPRM1	Alcohol	Promoter	Hypermethylation	NA	Human/blood	Bisulfite sequencing	Zhang et al. (2012)
OPRM1	Opioids	Promoter	Hypermethylation	NA	Human/blood, sperm	Pyrosequencing; bisulfite sequencing	Chorbov, Todorov, Lynskey, and Cicero (2011) and Nielsen et al. (2009)
PDYN	Alcohol	3' UTR CpG-SNP	Hypermethylation	mRNA increase	Human/prefrontal cortex	Pyrosequencing	Taqi et al. (2011)
PP1c	Cocaine	Promoter	Hypermethylation	mRNA decrease	Mouse/nucleus accumbens	MeDIP/methylation-specific qPCR	Anier et al. (2010)
PPM1G	Alcohol	3' of gene sequence	Hypermethylation	mRNA decrease	Human/blood	Methylation array, Sequenom	Ruggeri et al. (2015)
SNCA	Alcohol	Promoter	Hypermethylation	Both decrease	Human/blood	Methylation-specific qPCR	Bonsch et al. (2005)
Sox10	Cocaine	Promoter	Hypomethylation	NA	Rat/white matter	Bisulfite sequencing	Nielsen, Huang, et al. (2012)
Sty2	Alcohol	First exon	Hypermethylation	mRNA decrease	Rat/prefrontal cortex	Pyrosequencing	Barbier et al. (2015)

MeDIP, Methylated DNA immune-precipitation; *NA*, not available; *qPCR*, quantitative PCR; *SNP*, single-nucleotide polymorphism; *UTR*, untranslated region.

Studies of ethanol

The number of studies focused on DNA methylation changes in ethanol addiction has grown in recent years (Krishnan, Sakharkar, Teppen, Berkel, & Pandey, 2014; Kyzar & Pandey, 2015; Ponomarev, 2013). Many such studies come from human subjects, particularly from peripheral blood samples. This work has provided valuable information on ethanol action.

One unique feature of ethanol, with respect to DNA methylation, is that ethanol metabolism directly affects methyl donors throughout the body. For example, chronic ethanol exposure can induce a folate deficiency, resulting in decreased levels of S-adenosylmethionine (SAM) and increased levels of its precursor homocysteine (Blasco et al., 2005; Hamid, Wani, & Kaur, 2009). As SAM is the primary methyl donor for most biological reactions, this effect of ethanol would be expected to decrease DNA methylation. In addition, the ethanol metabolite, acetaldehyde, was shown to inhibit DNMT activity in vitro (Garro, McBeth, Lima, & Lieber, 1991). Decreased expression of *Dnmt3a* and *Dnmt3b*, but not *Dnmt1*, was observed in whole blood of alcoholics, with a significant negative correlation between *Dnmt3b* expression and blood ethanol concentrations (Bonsch et al., 2006). Although the effect of ethanol on DNMT expression in adult brain is still unclear, a human study using microarray technology identified a 20–30% downregulation of *Dnmt1* across the superior frontal cortex and amygdala of alcoholics (Ponomarev, Wang, Zhang, Harris, & Mayfield, 2012). In parallel, in adolescent rat brain, it was recently reported that acute ethanol exposure significantly inhibited DNMT activity in the amygdala and bed nucleus of stria terminalis (BNST). However, this same study also demonstrated contradicting increases in *Dnmt1* and *Dnmt3a* mRNAs in the BNST together with no mRNA alterations in amygdala (Sakharkar et al., 2014). These findings indicate that DNMT enzymatic activity may not associate with transcriptional alterations. They also recommend further analyses, together with protein levels of DNMTs, in a defined brain region, cell type, and developmental stage. In fact, the same group carried out such an approach on cultured astrocytes and found that DNMT activity and DNMT3a protein expression were both attenuated by ethanol (X. Zhang et al., 2014), which may be involved in astrocyte-mediated inhibition of neuronal plasticity by ethanol exposure. Consistent with the aforementioned reduction of SAM after chronic ethanol administration, global DNA demethylation was observed in liver and colon of alcoholics (Choi et al., 1999; Lu et al., 2000). However, a significant increase (10%) of genomic DNA methylation in blood of patients with alcoholism was reported previously (Bonsch, Lenz, Reulbach, Kornhuber, & Bleich, 2004). This not only suggests ethanol-induced alteration in global DNA methylation is tissue specific and cell-type specific but also warrants future replication studies with the addition of mapping DNA methylation genome-wide.

Altered methylation of numerous genes, most derived from candidate gene approaches, has been associated with alcohol use disorders (Table 8.1). For example, the gene encoding the N-methyl-D-aspartate (NMDA) receptor NR2B underwent promoter demethylation after chronic ethanol treatment in primary mouse cortical neuronal cultures. This was associated with increased transcription of the gene (Marutha Ravindran & Ticku, 2005). Furthermore, a clinical study demonstrated that the degree of methylation of the *NR2B* gene promoter in peripheral blood during withdrawal negatively correlated with the severity of ethanol consumption (Biermann et al., 2009). Similarly, elevated DNA methylation of the *HERP* and *SNCA* genes associated with repression of gene transcription in peripheral blood of alcoholics (Bleich et al., 2006; Bonsch, Lenz, Kornhuber, & Bleich, 2005). Barbier et al. (2015) reported that in rats after 3 weeks of abstinence from chronic ethanol drinking, there was increased DNA methylation, and reduced expression, of several genes encoding synaptic proteins involved in neurotransmitter release in the medial prefrontal cortex (mPFC). They further showed that intra-mPFC infusion of RG108 blocked the increased ethanol consumption observed under these conditions along blockade of the downregulation of four of the seven regulated mRNA transcripts. Finally, viral-mediated suppression of one of these transcripts (*Syt2*) increased alcohol drinking, thus directly linking DNA methylation–regulated changes in gene transcription to alcohol addiction (Barbier et al., 2015).

However, this negative correlation between DNA methylation and gene expression is not always seen. By analyzing the *PDYN* gene in the dorsolateral PFC of human alcoholics postmortem, DNA methylation status was examined on CpG sites overlapping with single-nucleotide polymorphisms (SNPs) known to be associated with alcohol dependence. Interestingly, a methylation increase at such CpG-SNP sites within the 3′ untranslated region (3′ UTR) was observed and positively correlated with *Pdyn* transcription and with increased vulnerability to alcohol dependence (Taqi et al., 2011). This observation further emphasizes that the effects of DNA methylation on gene transcription are complex and region specific.

The differential methylation of certain genes observed in peripheral blood suggests that such epigenetic profiles might serve as a valid biomarker for alcohol use disorders. Although further work is needed to determine how DNA methylation patterns in blood compare with those in several relevant brain regions, the same genes need not be influenced in blood and brain for peripheral DNA methylation measures to serve an important biomarker function. To better account for DNA methylation changes associated with alcohol use disorders, a recent consortium carried out a genome-wide analysis of DNA methylation in peripheral blood of 18 monozygotic twin pairs discordant for alcohol use disorders and identified 77 differentially methylated regions most of which (68%) were hypermethylated (Ruggeri et al., 2015). The 3′-*PPM1G* (protein phosphatase 1G) gene locus displayed the most significant association with alcohol use disorders in the genome-wide analysis and in a replication cohort (Ruggeri et al., 2015). Moreover, *PPM1G* hypermethylation was associated with increased activation of the subthalamic nucleus as measured by functional brain imaging, early escalation of alcohol use, and increased impulsiveness. This study illustrates the power of genome-wide approaches, as discussed in more detail later.

METHYL-CYTOSINE OXIDATION AND ITS TET CATALYZING ENZYMES IN ADDICTION

Compared to most other epigenetic mechanisms, DNA methylation was initially believed to mediate long-term gene silencing. This view was supported by the fact that no definitive DNA demethylation machinery was known (Ooi & Bestor, 2008). Moreover, as DNA methylation patterns are largely established and maintained during DNA replication, it was unclear how dynamic is DNA methylation in nondividing neurons. Despite these earlier views, newer evidence indicated that DNA methylation turnover does occur in the fully differentiated adult brain (Feng et al., 2010; Ma et al., 2009; Miller & Sweatt, 2007), which suggested the existence of DNA demethylation mechanisms. Toward the end of 2009, two groups independently demonstrated that Ten-eleven translocation (TET) protein 1 oxidizes 5-methylcytosine (5mC) into 5-hydroxymethylcytosine (5hmC) (Kriaucionis & Heintz, 2009; Tahiliani et al., 2009). Two other members of the same family, TET2 and TET3, were subsequently shown to possess similar enzymatic activity. It was also found that 5hmC can be further oxidized into 5-formylcytosine (5fC) and 5-carboxylcytosine (5caC) successively (He et al., 2011; Ito et al., 2011). These findings supported the scheme that 5mC oxidation leads to DNA demethylation. Indeed, several biochemical pathways, such as the thymine-DNA glycosylase (TDG) base excision repair pathway, have been implicated in promoting the conversion of 5mC to 5hmC and subsequently to DNA demethylation (Branco, Ficz, & Reik, 2012; Pastor, Aravind, & Rao, 2013; Wu & Zhang, 2011). In addition, increasing evidence has shown that the 5hmC modification itself serves a transcriptional regulatory role, promoting either transcription activation or repression through multiple potential pathways (Pastor et al., 2013). Given the fact that TET-mediated 5mC oxidation can occur independently of DNA replication, these discoveries provide mechanistic details for the dynamic demethylation of DNA in postmitotic neurons.

One striking feature of 5hmC is that it is most enriched in brain compared with other organs (Globisch et al., 2010; Kriaucionis & Heintz, 2009; Szwagierczak, Bultmann, Schmidt, Spada, & Leonhardt, 2010). Indeed, TET enzymes and 5hmC have been shown to play important roles in active DNA demethylation in hippocampus where they have been implicated in neural development, aging, and learning and memory (Guo, Su, Zhong, Ming, & Song, 2011; Kaas et al., 2013; Li et al., 2014; Rudenko et al., 2013; Szulwach et al., 2011; R. R. Zhang et al., 2013).

Support for a role of TET1 and 5hmC in drug addiction comes from a study by our group (Feng et al., 2015). We found that chronic cocaine administration, by repeated IP injections, decreased TET1, but not TET2 or TET3, in the mouse NAc, an effect observed as well in this region of cocaine addicts examined postmortem. To understand the functional effect of TET1 on cocaine action, we virally knocked down or overexpressed TET1 in the adult NAc and showed that TET1 negatively regulates drug reward behavior. These results suggest that the cocaine-induced downregulation of TET1 in NAc serves to increase cocaine action. As TET1 oxidizes 5mC into 5hmC, we also performed global measurement of both modifications in NAc and did not detect any regulation by cocaine administration. This finding suggests that any 5hmC changes induced by cocaine in this brain region may be locus specific, driving the genome-wide mapping of 5hmC discussed in the next section.

Although little work has been performed on TET or 5hmC in other addiction paradigms, it was shown recently that TET1 mRNA levels are increased in the PFC of psychotic patients that were chronic alcohol abusers (Guidotti et al., 2013). Further studies of TET proteins and 5hmC in rodent ethanol models are now warranted to study the functional ramifications of the human findings.

There have also been reports for other mechanisms of active DNA demethylation, in particular, a role for growth arrest and DNA damage 45 (GADD45) protein family members, which have been shown to mediate DNA demethylation during cell differentiation and cellular stress responses (Ma et al., 2009; Niehrs & Schafer, 2012). It was reported that knockdown of GADD45β blocks the ability of NMDA receptor activation to reduce methylation of the *Bdnf* gene promoter, and induce *Bdnf* mRNA levels, in cultured neurons (Gavin et al., 2015; Ma et al., 2009). This important finding establishes the involvement of GADD45 in gene regulation in brain. Moreover, given the important influence of BDNF signaling in drug addiction, work is now needed to study a possible role for GADD45 in addiction-related abnormalities. Indeed, Koo et al. (2012) demonstrated that chronic morphine administration decreases GADD45γ in the NAc and that overexpression of the protein in this region promotes rewarding responses to the drug. Any relationship between these findings and altered DNA methylation remains to be established.

GENOME-WIDE MAPPING OF DNA MODIFICATIONS IN ADDICTION

Although most studies to date have taken a candidate gene approach to identify genes regulated by DNA methylation in addiction models (Table 8.1), it is essential to gather an unbiased, genome-wide view of such regulation. We and other groups have used chromatin immunoprecipitation (ChIP) followed by promoter arrays and more recently ChIP sequencing (seq) to map drug-induced changes in several histone modifications in specific brain regions (Renthal et al., 2009; Zhou, Yuan, Mash, & Goldman, 2011), and similar approaches are now a high priority for DNA methylation.

Given the role of TET1 in NAc in cocaine action as already described, we mapped cocaine-induced 5hmC alterations in this brain region. By use of a chemical biotin labeling approach (Song et al., 2011), 5hmC-enriched DNA fragments were purified from mouse NAc and sequenced (Feng et al., 2015). In total, we recognized more than 20,000 differential regions, the majority of which occurred in gene body and intergenic regions. To better understand the potential function of 5hmC regulation in intergenic regions, we focused on 5hmC dynamics at putative enhancer regions. Enhancers are regulatory elements that can exist long distances from transcription start sites. We generated ChIP-seq maps for two histone marks, H3K4me1 and H3K27ac, both of which have been used to identify enhancers (Creyghton et al., 2010). We then used a combinatorial approach to define distinct chromatin states (Ernst et al., 2011) at nonpromoter regions based on altered binding of these histone marks plus 5hmC. We observed dynamic regulation of chromatin states at putative enhancers in response to cocaine (Feng et al., 2015). Moreover, we detected cocaine regulation of 5hmC around exon boundaries, sites correlated positively with alternative splicing changes detected by RNA-seq. Overlay of RNA-seq with 5hmC data in coding regions further revealed that increased levels of 5hmC in gene bodies is associated positively with increased steady state transcription of that gene or its greater inducibility in response to a subsequent cocaine challenge (Feng et al., 2015). The genes that displayed this regulation are highly enriched in addiction-related gene categories. Moreover, we observed that, at least at some loci, the cocaine-induced changes in 5hmC persist at least 30 days after cocaine administration (Feng et al., 2015). It would be interesting to map 5hmC at this later time point as well as in cocaine self-administration models and to compare such patterns to those of 5mC. Studies of drug regulation of DNA methylation in other brain reward regions are also needed.

Another recent study demonstrated dynamic changes in DNA methylation in the NAc during incubation of cocaine craving in rats (Massart et al., 2015). By use of methyl-DNA immunoprecipitation—which is mostly selective for 5mC—followed by microarray analysis, the authors examined methylation changes at gene promoters of all coding genes and across the entire gene length of a custom panel of 47 candidate genes previously implicated in drug addiction. They identified broad and time-dependent alterations in DNA methylation in this brain region after cocaine withdrawal and cue-induced cocaine seeking. Interestingly, patterns of DNA methylation varied dramatically between 1 day and 30 days of withdrawal and were rapidly reversed within 1 h of cue-induced reinstatement (Massart et al., 2015). These findings demonstrate the highly dynamic regulation of the DNA methylome in brain and raise the possibility that certain alterations might offer prognostic value if detectable in peripheral tissues.

Studies of the DNA methylome in alcoholism are also beginning to appear. In an investigation on frontal cortex of human alcoholics by use of NimbleGen Human DNA Methylation promoter arrays, numerous differential methylation loci were observed in both novel and known target genes that are either hypo- or hypermethylated in alcoholism (Manzardo, Henkhaus, & Butler, 2012). Another microarray-based genome-wide DNA methylation analysis, this one with Illumina chips, surveyed the DNA methylome in lymphocytes from ~60 alcohol-dependent patients with a similar number of controls (R. Zhang et al., 2013). The study found 1710 CpG sites that were differentially methylated between alcoholics and controls, with the majority hypomethylated in patients. The identified genes were enriched in several interesting categories such as stress, immune response, and signal transduction. Other Illumina chip–based studies also found DNA methylation differences in the blood or saliva of alcoholics versus control subjects (Harlaar et al., 2014; Philibert et al., 2014; H. Zhang et al., 2013). In an investigation of 18 monozygotic twin pairs discordant for alcohol use disorders, genome-wide DNA methylation array screening and Sequenom EpiTYPER validation of the same peripheral blood DNA samples identified ~20 differentially methylated regions associated with alcohol (Ruggeri et al., 2015). It would be important to validate these novel changes in methylation in additional cohorts and to determine whether similar changes occur within relevant brain reward regions.

Even though the vast majority of studies to date have focused on DNA methylation associated with coding genes, recent work suggests that addiction can also regulate DNA methylation at normally silenced repetitive elements. By applying a novel transcriptome analysis approach, namely, weighted gene coexpression network analysis (Zhang & Horvath, 2005), Ponomarev et al. (2012) studied transcriptional alterations associated with alcohol use disorders in postmortem human brain. The authors identified previously unrecognized epigenetic determinants of gene coexpression relationships and discovered novel markers of chromatin modifications in the amygdala and superior frontal cortex of alcoholics (Ponomarev et al., 2012). Higher expression levels of endogenous retroviruses in alcoholics were then confirmed to be associated with DNA hypomethylation, suggesting a critical role of DNA methylation in alcohol addiction. Consistent with this, chronic cocaine administration has been shown to trigger the loss of H3K9me3 at repetitive genomic sequences in mouse NAc and hence increase their expression (Maze et al., 2011). It would be interesting to follow up these observations with measures of DNA methylation at these loci after cocaine exposure. Although further work is needed, studies to date suggest that the control of repetitive elements by DNA methylation and related repressive chromatin mechanisms represents a novel form of epigenetic regulation in the addicted brain.

CHALLENGES AND FUTURE DIRECTIONS

One main obstacle for DNA epigenetic studies in addiction is how to integrate the various methodologies used (eg, Bock et al., 2010). In contrast to relatively standard ChIP-seq methods for other histone modifications, numerous approaches have been used to measure DNA epigenetic modifications. And with novel forms of DNA modifications discovered in the past few years (eg, 5hmC, 5fC, and 5caC), it would not be surprising to see still additional approaches being introduced (Booth et al., 2012; Song et al., 2013; Wu, Wu, Shen, & Zhang, 2014; Yu et al., 2012). Each currently used approach has its own pros and cons, which makes the choice of the right methodology particularly difficult.

For example, experimenters have to decide between cost and coverage, coding regions and intergenic regions, base resolution and fragmented resolution, quantitative and relative, small amount and large amount of starting DNA, and so on. Given the high cost and demanding bioinformatic support in whole-methylome analyses, initial studies largely depended on candidate gene approaches based on restriction enzyme cutting, antibody pulldown, or sodium bisulfite conversion. The inherent limitations of these approaches presumably explain why only relatively few genes were initially recognized to undergo methylation alterations in addiction models. The major imperative today, therefore, is to use next-generation sequencing technology with whole-genome coverage at single-base resolution, distinguishing between different cytosine modifications. This approach will become more feasible as sequencing costs decline. In the meantime, it is essential to cross-compare datasets derived not only from different brain regions, peripheral tissues, and addiction paradigms but also derived from various bioinformatic platforms (Maze et al., 2014). Similarly, an important goal is to relate DNA methylomes to variations in DNA sequence data as they relate to vulnerability to drug addiction (H. Zhang et al., 2014).

As DNA epigenetic changes are identified, it will become crucial to manipulate these epigenetic states at selective loci to obtain causal insight into their role in gene regulation. A study from our group, using engineered zinc finger proteins or transcription activator-like effectors to target single types of histone modifications to single genes within a single brain region of interest in vivo, provides a means of obtaining such causal data (Heller et al., 2014). The increasing availability of genome-editing tools (Tuesta & Zhang, 2014) offers additional technical approaches to achieve this important goal.

Another pressing question facing the field is whether DNA modifications, or any epigenetic modifications more broadly defined, are specific to a given drug of abuse or specific brain region. The existing literature, although still limited, suggests some common actions as well as many distinct actions. For example, one study demonstrated that DNA methylation at selected genes in VTA is required for the formation of reward-related memories, effects not seen for the NAc (Day et al., 2013).

We have been using the term epigenetics to refer to any chromatin modification that controls genomic function. However, the term is also used to describe heritable changes that occur without alterations of the underlying DNA sequence. Studies suggest that this heritable form of epigenetic regulation may also be involved in addiction. Several studies have shown that exposure of male rodent to drugs of abuse, or to stress, triggers behavioral changes in offspring (eg, Bale, 2015; Dietz et al., 2011; Gapp et al., 2014; Szutorisz et al., 2014). For example, prior cocaine self-administration by fathers can affect their offspring's cocaine acquisition behavior (Vassoler, White, Schmidt, Sadri-Vakili, & Pierce, 2013). DNA methylation has been implicated in such inheritance of behavioral experience: sperm DNA from F0 males exposed to odor fear conditioning and F1 naive offspring revealed CpG hypomethylation of the *Olfr151* gene, which was associated with increased behavioral sensitivity (Dias & Ressler, 2014). In addition, in vitro fertilization, F2 inheritance, and cross-fostering revealed that these transgenerational effects are inherited via parental gametes. Whether drugs of abuse can also affect transgenerational responses through DNA epigenetic modifications remains unknown. One study that probed this question has indeed demonstrated dynamic DNA methylation by reduced representation bisulfite sequencing DNA methylation profiling. The authors compared the NAc methylome in animals with and without parental cannabinoid exposure (Watson et al., 2015) and identified 1027 differentially methylated regions in the NAc of F1 adults associated with parental cannabinoid exposure. Many of the regions were related to genes involved in glutamatergic synaptic regulation. The goal of future studies is to provide causal evidence and precise molecular mechanisms by which such regulation occurs.

ACKNOWLEDGMENTS

Preparation of this review was supported by grants from the National Institute on Drug Abuse. The authors apologize for work not cited in this review due to space limitations.

REFERENCES

Anier, K., Malinovskaja, K., Aonurm-Helm, A., Zharkovsky, A., & Kalda, A. (2010). DNA methylation regulates cocaine-induced behavioral sensitization in mice. *Neuropsychopharmacology: Official Publication of the American College of Neuropsychopharmacology, 35*(12), 2450–2461. http://dx.doi.org/10.1038/npp.2010.128, pii:npp2010128.
Anthony, J. C., Warner, L. A., & Kessler, R. C. (1994). Comparative epidemiology of dependence on tobacco, alcohol, controlled substances, and inhalants: basic findings from the National Comorbidity Survey. *Experimental and Clinical Psychopharmacology, 2*(3), 244–268.
Bale, T. L. (2015). Epigenetic and transgenerational reprogramming of brain development. *Nature Reviews Neuroscience, 16*(6), 332–344. http://dx.doi.org/10.1038/nrn3818.
Barbier, E., Tapocik, J. D., Juergens, N., Pitcairn, C., Borich, A., Schank, J. R., et al. (2015). DNA methylation in the medial prefrontal cortex regulates alcohol-induced behavior and plasticity. *Journal of Neuroscience, 35*(15), 6153–6164. http://dx.doi.org/10.1523/JNEUROSCI.4571-14.2015.
Biermann, T., Reulbach, U., Lenz, B., Frieling, H., Muschler, M., Hillemacher, T., et al. (2009). N-methyl-D-aspartate 2b receptor subtype (NR2B) promoter methylation in patients during alcohol withdrawal. *Journal of Neural Transmission, 116*(5), 615–622. http://dx.doi.org/10.1007/s00702-009-0212-2.
Blasco, C., Caballeria, J., Deulofeu, R., Lligona, A., Pares, A., Lluis, J. M., et al. (2005). Prevalence and mechanisms of hyperhomocysteinemia in chronic alcoholics. *Alcoholism, Clinical and Experimental Research, 29*(6), 1044–1048.
Bleich, S., Lenz, B., Ziegenbein, M., Beutler, S., Frieling, H., Kornhuber, J., et al. (2006). Epigenetic DNA hypermethylation of the HERP gene promoter induces down-regulation of its mRNA expression in patients with alcohol dependence. *Alcoholism, Clinical and Experimental Research, 30*(4), 587–591. http://dx.doi.org/10.1111/j.1530-0277.2006.00068.x.
Bock, C., Tomazou, E. M., Brinkman, A. B., Muller, F., Simmer, F., Gu, H., et al. (2010). Quantitative comparison of genome-wide DNA methylation mapping technologies. *Nature Biotechnology, 28*(10), 1106–1114. http://dx.doi.org/10.1038/nbt.1681.
Bonsch, D., Lenz, B., Fiszer, R., Frieling, H., Kornhuber, J., & Bleich, S. (2006). Lowered DNA methyltransferase (DNMT-3b) mRNA expression is associated with genomic DNA hypermethylation in patients with chronic alcoholism. *Journal of Neural Transmission, 113*(9), 1299–1304. http://dx.doi.org/10.1007/s00702-005-0413-2.

Bonsch, D., Lenz, B., Kornhuber, J., & Bleich, S. (2005). DNA hypermethylation of the alpha synuclein promoter in patients with alcoholism. *Neuroreport, 16*(2), 167–170.

Bonsch, D., Lenz, B., Reulbach, U., Kornhuber, J., & Bleich, S. (2004). Homocysteine associated genomic DNA hypermethylation in patients with chronic alcoholism. *Journal of Neural Transmission, 111*(12), 1611–1616. http://dx.doi.org/10.1007/s00702-004-0232-x.

Booth, M. J., Branco, M. R., Ficz, G., Oxley, D., Krueger, F., Reik, W., et al. (2012). Quantitative sequencing of 5-methylcytosine and 5-hydroxymethylcytosine at single-base resolution. *Science, 336*(6083), 934–937. http://dx.doi.org/10.1126/science.1220671.

Branco, M. R., Ficz, G., & Reik, W. (2012). Uncovering the role of 5-hydroxymethylcytosine in the epigenome. *Nature Reviews Genetics, 13*(1), 7–13. http://dx.doi.org/10.1038/nrg3080 pii:nrg3080.

Choi, S. W., Stickel, F., Baik, H. W., Kim, Y. I., Seitz, H. K., & Mason, J. B. (1999). Chronic alcohol consumption induces genomic but not p53-specific DNA hypomethylation in rat colon. *Journal of Nutrition, 129*(11), 1945–1950.

Chorbov, V. M., Todorov, A. A., Lynskey, M. T., & Cicero, T. J. (2011). Elevated levels of DNA methylation at the OPRM1 promoter in blood and sperm from male opioid addicts. *Journal of Opioid Management, 7*(4), 258–264.

Creyghton, M. P., Cheng, A. W., Welstead, G. G., Kooistra, T., Carey, B. W., Steine, E. J., et al. (2010). Histone H3K27ac separates active from poised enhancers and predicts developmental state. *Proceedings of the National Academy of Sciences of the United States of America, 107*(50), 21931–21936. http://dx.doi.org/10.1073/pnas.1016071107.

Day, J. J., Childs, D., Guzman-Karlsson, M. C., Kibe, M., Moulden, J., Song, E., et al. (2013). DNA methylation regulates associative reward learning. *Nature Neuroscience, 16*(10), 1445–1452. http://dx.doi.org/10.1038/nn.3504.

Day, J. J., & Sweatt, J. D. (2011). Epigenetic mechanisms in cognition. *Neuron, 70*(5), 813–829. http://dx.doi.org/10.1016/j.neuron.2011.05.019.

Deng, J. V., Rodriguiz, R. M., Hutchinson, A. N., Kim, I. H., Wetsel, W. C., & West, A. E. (2010). MeCP2 in the nucleus accumbens contributes to neural and behavioral responses to psychostimulants. *Nature Neuroscience, 13*(9), 1128–1136. http://dx.doi.org/10.1038/nn.2614, pii:nn.2614.

Deng, J. V., Wan, Y., Wang, X., Cohen, S., Wetsel, W. C., Greenberg, M. E., et al. (2014). MeCP2 phosphorylation limits psychostimulant-induced behavioral and neuronal plasticity. *Journal of Neuroscience, 34*(13), 4519–4527. http://dx.doi.org/10.1523/JNEUROSCI.2821-13.2014.

Dias, B. G., & Ressler, K. J. (2014). Parental olfactory experience influences behavior and neural structure in subsequent generations. *Nature Neuroscience, 17*(1), 89–96. http://dx.doi.org/10.1038/nn.3594.

Dietz, D. M., Laplant, Q., Watts, E. L., Hodes, G. E., Russo, S. J., Feng, J., et al. (2011). Paternal transmission of stress-induced pathologies. *Biological Psychiatry, 70*(5), 408–414. http://dx.doi.org/10.1016/j.biopsych.2011.05.005.

Ernst, J., Kheradpour, P., Mikkelsen, T. S., Shoresh, N., Ward, L. D., Epstein, C. B., et al. (2011). Mapping and analysis of chromatin state dynamics in nine human cell types. *Nature, 473*(7345), 43–49. http://dx.doi.org/10.1038/nature09906.

Feng, J., & Nestler, E. J. (2013). Epigenetic mechanisms of drug addiction. *Current Opinion in Neurobiology*. http://dx.doi.org/10.1016/j.conb.2013.01.001.

Feng, J., Shao, N., Szulwach, K. E., Vialou, V., Huynh, J., Zhong, C., et al. (2015). Role of Tet1 and 5-hydroxymethylcytosine in cocaine action. *Nature Neuroscience, 18*(4), 536–544. http://dx.doi.org/10.1038/nn.3976.

Feng, J., Zhou, Y., Campbell, S. L., Le, T., Li, E., Sweatt, J. D., et al. (2010). Dnmt1 and Dnmt3a maintain DNA methylation and regulate synaptic function in adult forebrain neurons. *Nature Neuroscience, 13*(4), 423–430. http://dx.doi.org/10.1038/nn.2514.

Gapp, K., Jawaid, A., Sarkies, P., Bohacek, J., Pelczar, P., Prados, J., et al. (2014). Implication of sperm RNAs in transgenerational inheritance of the effects of early trauma in mice. *Nature Neuroscience, 17*(5), 667–669. http://dx.doi.org/10.1038/nn.3695.

Garro, A. J., McBeth, D. L., Lima, V., & Lieber, C. S. (1991). Ethanol consumption inhibits fetal DNA methylation in mice: implications for the fetal alcohol syndrome. *Alcoholism, Clinical and Experimental Research, 15*(3), 395–398.

Gavin, D. P., Kusumo, H., Sharma, R. P., Guizzetti, M., Guidotti, A., & Pandey, S. C. (2015). Gadd45b and N-methyl-D-aspartate induced DNA demethylation in postmitotic neurons. *Epigenomics*, 7(4), 567–579. http://dx.doi.org/10.2217/epi.15.12.

Globisch, D., Munzel, M., Muller, M., Michalakis, S., Wagner, M., Koch, S., et al. (2010). Tissue distribution of 5-hydroxymethylcytosine and search for active demethylation intermediates. *PLoS One*, 5(12), e15367. http://dx.doi.org/10.1371/journal.pone.0015367.

Godino, A., Jayanthi, S., & Cadet, J. L. (2015). Epigenetic landscape of amphetamine and methamphetamine addiction in rodents. *Epigenetics*, 10(7), 574–580. http://dx.doi.org/10.1080/15592294.2015.1055441.

Guidotti, A., Dong, E., Gavin, D. P., Veldic, M., Zhao, W., Bhaumik, D. K., et al. (2013). DNA methylation/demethylation network expression in psychotic patients with a history of alcohol abuse. *Alcoholism, Clinical and Experimental Research*, 37(3), 417–424. http://dx.doi.org/10.1111/j.1530-0277.2012.01947.x.

Guo, J. U., Su, Y., Zhong, C., Ming, G. L., & Song, H. (2011). Hydroxylation of 5-methylcytosine by TET1 promotes active DNA demethylation in the adult brain. *Cell*, 145(3), 423–434. http://dx.doi.org/10.1016/j.cell.2011.03.022, pii:S0092-8674(11)00299-6.

Hamid, A., Wani, N. A., & Kaur, J. (2009). New perspectives on folate transport in relation to alcoholism-induced folate malabsorption–association with epigenome stability and cancer development. *FEBS Journal*, 276(8), 2175–2191. http://dx.doi.org/10.1111/j.1742-4658.2009.06959.x.

Harlaar, N., Bryan, A. D., Thayer, R. E., Karoly, H. C., Oien, N., & Hutchison, K. E. (2014). Methylation of a CpG site near the ALDH1A2 gene is associated with loss of control over drinking and related phenotypes. *Alcoholism, Clinical and Experimental Research*, 38(3), 713–721. http://dx.doi.org/10.1111/acer.12312.

He, Y. F., Li, B. Z., Li, Z., Liu, P., Wang, Y., Tang, Q., et al. (2011). Tet-mediated formation of 5-carboxylcytosine and its excision by TDG in mammalian DNA. *Science*, 333(6047), 1303–1307. http://dx.doi.org/10.1126/science.1210944, pii:science.1210944.

Heberlein, A., Muschler, M., Frieling, H., Behr, M., Eberlein, C., Wilhelm, J., et al. (2013). Epigenetic down regulation of nerve growth factor during alcohol withdrawal. *Addiction Biology*, 18(3), 508–510. http://dx.doi.org/10.1111/j.1369-1600.2010.00307.x.

Heller, E. A., Cates, H. M., Pena, C. J., Sun, H., Shao, N., Feng, J., et al. (2014). Locus-specific epigenetic remodeling controls addiction- and depression-related behaviors. *Nature Neuroscience*, 17(12), 1720–1727. http://dx.doi.org/10.1038/nn.3871.

Hillemacher, T., Frieling, H., Hartl, T., Wilhelm, J., Kornhuber, J., & Bleich, S. (2009). Promoter specific methylation of the dopamine transporter gene is altered in alcohol dependence and associated with craving. *Journal of Psychiatric Research*, 43(4), 388–392. http://dx.doi.org/10.1016/j.jpsychires.2008.04.006.

Hillemacher, T., Frieling, H., Luber, K., Yazici, A., Muschler, M. A., Lenz, B., et al. (2009). Epigenetic regulation and gene expression of vasopressin and atrial natriuretic peptide in alcohol withdrawal. *Psychoneuroendocrinology*, 34(4), 555–560. http://dx.doi.org/10.1016/j.psyneuen.2008.10.019.

Hyman, S. E., Malenka, R. C., & Nestler, E. J. (2006). Neural mechanisms of addiction: the role of reward-related learning and memory. *Annual Review of Neuroscience*, 29, 565–598. http://dx.doi.org/10.1146/annurev.neuro.29.051605.113009.

Im, H. I., Hollander, J. A., Bali, P., & Kenny, P. J. (2010). MeCP2 controls BDNF expression and cocaine intake through homeostatic interactions with microRNA-212. *Nature Neuroscience*, 13(9), 1120–1127. http://dx.doi.org/10.1038/nn.2615, pii:nn.2615.

Ito, S., Shen, L., Dai, Q., Wu, S. C., Collins, L. B., Swenberg, J. A., et al. (2011). Tet proteins can convert 5-methylcytosine to 5-formylcytosine and 5-carboxylcytosine. *Science*, 333(6047), 1300–1303. http://dx.doi.org/10.1126/science.1210597, pii:science.1210597.

Jaenisch, R., & Bird, A. (2003). Epigenetic regulation of gene expression: how the genome integrates intrinsic and environmental signals. *Nature Genetics*, 33(Suppl.), 245–254.

Jayanthi, S., McCoy, M. T., Chen, B., Britt, J. P., Kourrich, S., Yau, H. J., et al. (2014). Methamphetamine downregulates striatal glutamate receptors via diverse epigenetic mechanisms. *Biological Psychiatry*, 76(1), 47–56. http://dx.doi.org/10.1016/j.biopsych.2013.09.034.

Kaas, G. A., Zhong, C., Eason, D. E., Ross, D. L., Vachhani, R. V., Ming, G. L., et al. (2013). TET1 controls CNS 5-methylcytosine hydroxylation, active DNA demethylation, gene transcription, and memory formation. *Neuron*, 79(6), 1086–1093. http://dx.doi.org/10.1016/j.neuron.2013.08.032, pii:S0896-6273(13)00791-5.

Kalivas, P. W., & Volkow, N. D. (2011). New medications for drug addiction hiding in glutamatergic neuro-plasticity. *Molecular Psychiatry*, *16*(10), 974–986. http://dx.doi.org/10.1038/mp.2011.46.

Kenny, P. J. (2014). Epigenetics, microRNA, and addiction. *Dialogues in Clinical Neuroscience*, *16*(3), 335–344.

Koo, J. W., Mazei-Robison, M. S., Chaudhury, D., Juarez, B., LaPlant, Q., Ferguson, D., et al. (2012). BDNF is a negative modulator of morphine action. *Science*, *338*(6103), 124–128. http://dx.doi.org/10.1126/science.1222265.

Kriaucionis, S., & Heintz, N. (2009). The nuclear DNA base 5-hydroxymethylcytosine is present in Purkinje neurons and the brain. *Science*, *324*(5929), 929–930.

Krishnan, H. R., Sakharkar, A. J., Teppen, T. L., Berkel, T. D., & Pandey, S. C. (2014). The epigenetic landscape of alcoholism. *International Review of Neurobiology*, *115*, 75–116. http://dx.doi.org/10.1016/B978-0-12-801311-3.00003-2.

Kyzar, E. J., & Pandey, S. C. (2015). Molecular mechanisms of synaptic remodeling in alcoholism. *Neuroscience Letters*. http://dx.doi.org/10.1016/j.neulet.2015.01.051.

LaPlant, Q., & Nestler, E. J. (2011). CRACKing the histone code: cocaine's effects on chromatin structure and function. *Hormones and Behavior*, *59*(3), 321–330. http://dx.doi.org/10.1016/j.yhbeh.2010.05.015, pii:S0018-506X(10)00158-3.

LaPlant, Q., Vialou, V., Covington, H. E., 3rd, Dumitriu, D., Feng, J., Warren, B. L., et al. (2010). Dnmt3a regulates emotional behavior and spine plasticity in the nucleus accumbens. *Nature Neuroscience*, *13*(9), 1137–1143. http://dx.doi.org/10.1038/nn.2619, pii:nn.2619.

Li, X., Wei, W., Zhao, Q. Y., Widagdo, J., Baker-Andresen, D., Flavell, C. R., et al. (2014). Neocortical Tet3-mediated accumulation of 5-hydroxymethylcytosine promotes rapid behavioral adaptation. *Proceedings of the National Academy of Sciences of the United States of America*, *111*(19), 7120–7125. http://dx.doi.org/10.1073/pnas.1318906111.

Lu, S. C., Huang, Z. Z., Yang, H., Mato, J. M., Avila, M. A., & Tsukamoto, H. (2000). Changes in methionine adenosyltransferase and S-adenosylmethionine homeostasis in alcoholic rat liver. *American Journal of Physiology Gastrointestinal and Liver Physiology*, *279*(1), G178–G185.

Lubin, F. D., Gupta, S., Parrish, R. R., Grissom, N. M., & Davis, R. L. (2011). Epigenetic mechanisms: critical contributors to long-term memory formation. *Neuroscientist*, *17*(6), 616–632. http://dx.doi.org/10.1177/1073858411386967.

Ma, D. K., Jang, M. H., Guo, J. U., Kitabatake, Y., Chang, M. L., Pow-Anpongkul, N., et al. (2009). Neuronal activity-induced Gadd45b promotes epigenetic DNA demethylation and adult neurogenesis. *Science*, *323*(5917), 1074–1077.

Manzardo, A. M., Henkhaus, R. S., & Butler, M. G. (2012). Global DNA promoter methylation in frontal cortex of alcoholics and controls. *Gene*, *498*(1), 5–12. http://dx.doi.org/10.1016/j.gene.2012.01.096.

Marutha Ravindran, C. R., & Ticku, M. K. (2005). Role of CpG islands in the up-regulation of NMDA receptor NR2B gene expression following chronic ethanol treatment of cultured cortical neurons of mice. *Neurochemistry International*, *46*(4), 313–327. http://dx.doi.org/10.1016/j.neuint.2004.10.004.

Massart, R., Barnea, R., Dikshtein, Y., Suderman, M., Meir, O., Hallett, M., et al. (2015). Role of DNA methylation in the nucleus accumbens in incubation of cocaine craving. *Journal of Neuroscience*, *35*(21), 8042–8058. http://dx.doi.org/10.1523/JNEUROSCI.3053-14.2015.

Maze, I., Feng, J., Wilkinson, M. B., Sun, H., Shen, L., & Nestler, E. J. (2011). Cocaine dynamically regulates heterochromatin and repetitive element unsilencing in nucleus accumbens. *Proceedings of the National Academy of Sciences of the United States of America*, *108*(7), 3035–3040. http://dx.doi.org/10.1073/pnas.1015483108, pii:1015483108.

Maze, I., & Nestler, E. J. (2011). The epigenetic landscape of addiction. *Annals of the New York Academy of Sciences*, *1216*, 99–113. http://dx.doi.org/10.1111/j.1749-6632.2010.05893.x.

Maze, I., Shen, L., Zhang, B., Garcia, B. A., Shao, N., Mitchell, A., et al. (2014). Analytical tools and current challenges in the modern era of neuroepigenomics. *Nature Neuroscience*, *17*(11), 1476–1490. http://dx.doi.org/10.1038/nn.3816.

Mikaelsson, M. A., & Miller, C. A. (2011). The path to epigenetic treatment of memory disorders. *Neurobiology of Learning and Memory*, *96*(1), 13–18. http://dx.doi.org/10.1016/j.nlm.2011.02.003.

Miller, C. A., & Sweatt, J. D. (2007). Covalent modification of DNA regulates memory formation. *Neuron*, *53*(6), 857–869.

Moore, L. D., Le, T., & Fan, G. (2013). DNA methylation and its basic function. *Neuropsychopharmacology: Official Publication of the American College of Neuropsychopharmacology, 38*(1), 23–38. http://dx.doi.org/10.1038/npp.2012.112.

Nelson, E. D., & Monteggia, L. M. (2011). Epigenetics in the mature mammalian brain: effects on behavior and synaptic transmission. *Neurobiology of Learning and Memory, 96*(1), 53–60. http://dx.doi.org/10.1016/j.nlm.2011.02.015.

Nestler, E. J. (2001). Molecular basis of long-term plasticity underlying addiction. *Nature Reviews Neuroscience, 2*(2), 119–128. http://dx.doi.org/10.1038/35053570.

Niehrs, C., & Schafer, A. (2012). Active DNA demethylation by Gadd45 and DNA repair. *Trends in Cell Biology, 22*(4), 220–227. http://dx.doi.org/10.1016/j.tcb.2012.01.002.

Nielsen, D. A., Huang, W., Hamon, S. C., Maili, L., Witkin, B. M., Fox, R. G., et al. (2012). Forced abstinence from cocaine self-administration is associated with DNA methylation changes in myelin genes in the corpus callosum: a preliminary study. *Frontiers in Psychiatry, 3*, 60. http://dx.doi.org/10.3389/fpsyt.2012.00060.

Nielsen, D. A., Utrankar, A., Reyes, J. A., Simons, D. D., & Kosten, T. R. (2012). Epigenetics of drug abuse: predisposition or response. *Pharmacogenomics, 13*(10), 1149–1160. http://dx.doi.org/10.2217/pgs.12.94.

Nielsen, D. A., Yuferov, V., Hamon, S., Jackson, C., Ho, A., Ott, J., et al. (2009). Increased OPRM1 DNA methylation in lymphocytes of methadone-maintained former heroin addicts. *Neuropsychopharmacology: Official Publication of the American College of Neuropsychopharmacology, 34*(4), 867–873. http://dx.doi.org/10.1038/npp.2008.108 pii:npp2008108.

Numachi, Y., Shen, H., Yoshida, S., Fujiyama, K., Toda, S., Matsuoka, H., et al. (2007). Methamphetamine alters expression of DNA methyltransferase 1 mRNA in rat brain. *Neuroscience Letters, 414*(3), 213–217. http://dx.doi.org/10.1016/j.neulet.2006.12.052.

Ooi, S. K., & Bestor, T. H. (2008). The colorful history of active DNA demethylation. *Cell, 133*(7), 1145–1148.

Ouko, L. A., Shantikumar, K., Knezovich, J., Haycock, P., Schnugh, D. J., & Ramsay, M. (2009). Effect of alcohol consumption on CpG methylation in the differentially methylated regions of H19 and IG-DMR in male gametes: implications for fetal alcohol spectrum disorders. *Alcoholism, Clinical and Experimental Research, 33*(9), 1615–1627. http://dx.doi.org/10.1111/j.1530-0277.2009.00993.x.

Pastor, W. A., Aravind, L., & Rao, A. (2013). TETonic shift: biological roles of TET proteins in DNA demethylation and transcription. *Nature Reviews Molecular Cell Biology, 14*(6), 341–356. http://dx.doi.org/10.1038/nrm3589, pii:nrm3589.

Philibert, R. A., Gunter, T. D., Beach, S. R., Brody, G. H., & Madan, A. (2008). MAOA methylation is associated with nicotine and alcohol dependence in women. *American Journal of Medical Genetics Part B Neuropsychiatric Genetics: The Official Publication of the International Society of Psychiatric Genetics, 147B*(5), 565–570. http://dx.doi.org/10.1002/ajmg.b.30778.

Philibert, R. A., Penaluna, B., White, T., Shires, S., Gunter, T., Liesveld, J., et al. (2014). A pilot examination of the genome-wide DNA methylation signatures of subjects entering and exiting short-term alcohol dependence treatment programs. *Epigenetics, 9*(9), 1212–1219. http://dx.doi.org/10.4161/epi.32252.

Ponomarev, I. (2013). Epigenetic control of gene expression in the alcoholic brain. *Alcohol Research, 35*(1), 69–76.

Ponomarev, I., Wang, S., Zhang, L., Harris, R. A., & Mayfield, R. D. (2012). Gene coexpression networks in human brain identify epigenetic modifications in alcohol dependence. *Journal of Neuroscience, 32*(5), 1884–1897. http://dx.doi.org/10.1523/JNEUROSCI.3136-11.2012.

Renthal, W., Kumar, A., Xiao, G., Wilkinson, M., Covington, H. E., 3rd, Maze, I., et al. (2009). Genome-wide analysis of chromatin regulation by cocaine reveals a role for sirtuins. *Neuron, 62*(3), 335–348. http://dx.doi.org/10.1016/j.neuron.2009.03.026 pii:S0896-6273(09)00241-4.

Robison, A. J., & Nestler, E. J. (2011). Transcriptional and epigenetic mechanisms of addiction. *Nature Reviews Neuroscience, 12*(11), 623–637. http://dx.doi.org/10.1038/nrn3111, pii:nrn3111.

Rogge, G. A., & Wood, M. A. (2013). The role of histone acetylation in cocaine-induced neural plasticity and behavior. *Neuropsychopharmacology: Official Publication of the American College of Neuropsychopharmacology, 38*(1), 94–110. http://dx.doi.org/10.1038/npp.2012.154.

Rudenko, A., Dawlaty, M. M., Seo, J., Cheng, A. W., Meng, J., Le, T., et al. (2013). Tet1 is critical for neuronal activity-regulated gene expression and memory extinction. *Neuron, 79*(6), 1109–1122. http://dx.doi.org/10.1016/j.neuron.2013.08.003, pii:S0896-6273(13)00714-9.

Ruggeri, B., Nymberg, C., Vuoksimaa, E., Lourdusamy, A., Wong, C. P., Carvalho, F. M., et al. (2015). Association of protein phosphatase PPM1G with alcohol use disorder and brain activity during behavioral control in a genome-wide methylation analysis. *American Journal of Psychiatry*, *172*(6), 543–552. http://dx.doi.org/10.1176/appi.ajp.2014.14030382.

Russo, S. J., & Nestler, E. J. (2013). The brain reward circuitry in mood disorders. *Nature Reviews Neuroscience*, *14*(9), 609–625. http://dx.doi.org/10.1038/nrn3381.

Sakharkar, A. J., Tang, L., Zhang, H., Chen, Y., Grayson, D. R., & Pandey, S. C. (2014). Effects of acute ethanol exposure on anxiety measures and epigenetic modifiers in the extended amygdala of adolescent rats. *International Journal of Neuropsychopharmacology*, *17*(12), 2057–2067. http://dx.doi.org/10.1017/S1461145714001047.

Schmidt, H. D., McGinty, J. F., West, A. E., & Sadri-Vakili, G. (2013). Epigenetics and psychostimulant addiction. *Cold Spring Harbor Perspectives in Medicine*, *3*(3). http://dx.doi.org/10.1101/cshperspect.a012047 pii:a012047.

Shin, J., Ming, G. L., & Song, H. (2014). DNA modifications in the mammalian brain. *Philosophical Transactions of the Royal Society of London Series B Biological Sciences*, *369*(1652). http://dx.doi.org/10.1098/rstb.2013.0512.

Song, C. X., Szulwach, K. E., Dai, Q., Fu, Y., Mao, S. Q., Lin, L., et al. (2013). Genome-wide profiling of 5-formylcytosine reveals its roles in epigenetic priming. *Cell*, *153*(3), 678–691. http://dx.doi.org/10.1016/j.cell.2013.04.001, pii:S0092-8674(13)00400-5.

Song, C. X., Szulwach, K. E., Fu, Y., Dai, Q., Yi, C., Li, X., et al. (2011). Selective chemical labeling reveals the genome-wide distribution of 5-hydroxymethylcytosine. *Nature Biotechnology*, *29*(1), 68–72. http://dx.doi.org/10.1038/nbt.1732.

Starkman, B. G., Sakharkar, A. J., & Pandey, S. C. (2012). Epigenetics-beyond the genome in alcoholism. *Alcohol Research*, *34*(3), 293–305.

Szulwach, K. E., Li, X., Li, Y., Song, C. X., Wu, H., Dai, Q., et al. (2011). 5-hmC-mediated epigenetic dynamics during postnatal neurodevelopment and aging. *Nature Neuroscience*, *14*(12), 1607–1616. http://dx.doi.org/10.1038/nn.2959.

Szutorisz, H., DiNieri, J. A., Sweet, E., Egervari, G., Michaelides, M., Carter, J. M., et al. (2014). Parental THC exposure leads to compulsive heroin-seeking and altered striatal synaptic plasticity in the subsequent generation. *Neuropsychopharmacology: Official Publication of the American College of Neuropsychopharmacology*, *39*(6), 1315–1323. http://dx.doi.org/10.1038/npp.2013.352.

Szwagierczak, A., Bultmann, S., Schmidt, C. S., Spada, F., & Leonhardt, H. (2010). Sensitive enzymatic quantification of 5-hydroxymethylcytosine in genomic DNA. *Nucleic Acids Research*, *38*(19), e181. http://dx.doi.org/10.1093/nar/gkq684.

Tahiliani, M., Koh, K. P., Shen, Y., Pastor, W. A., Bandukwala, H., Brudno, Y., et al. (2009). Conversion of 5-methylcytosine to 5-hydroxymethylcytosine in mammalian DNA by MLL partner TET1. *Science*, *324*(5929), 930–935.

Taqi, M. M., Bazov, I., Watanabe, H., Sheedy, D., Harper, C., Alkass, K., et al. (2011). Prodynorphin CpG-SNPs associated with alcohol dependence: elevated methylation in the brain of human alcoholics. *Addiction Biology*, *16*(3), 499–509. http://dx.doi.org/10.1111/j.1369-1600.2011.00323.x.

Tian, W., Zhao, M., Li, M., Song, T., Zhang, M., Quan, L., et al. (2012). Reversal of cocaine-conditioned place preference through methyl supplementation in mice: altering global DNA methylation in the prefrontal cortex. *PLoS One*, *7*(3), e33435. http://dx.doi.org/10.1371/journal.pone.0033435.

Tuesta, L. M., & Zhang, Y. (2014). Mechanisms of epigenetic memory and addiction. *EMBO Journal*, *33*(10), 1091–1103. http://dx.doi.org/10.1002/embj.201488106.

Vassoler, F. M., White, S. L., Schmidt, H. D., Sadri-Vakili, G., & Pierce, R. C. (2013). Epigenetic inheritance of a cocaine-resistance phenotype. *Nature Neuroscience*, *16*(1), 42–47. http://dx.doi.org/10.1038/nn.3280.

Walker, D. M., Cates, H. M., Heller, E. A., & Nestler, E. J. (2015). Regulation of chromatin states by drugs of abuse. *Current Opinion in Neurobiology*, *30*, 112–121. http://dx.doi.org/10.1016/j.conb.2014.11.002.

Watson, C. T., Szutorisz, H., Garg, P., Martin, Q., Landry, J. A., Sharp, A. J., et al. (2015). Genome-wide DNA methylation profiling reveals epigenetic changes in the rat nucleus accumbens associated with cross-generational effects of adolescent THC exposure. *Neuropsychopharmacology: Official Publication of the American College of Neuropsychopharmacology*. http://dx.doi.org/10.1038/npp.2015.155.

Wright, K. N., Hollis, F., Duclot, F., Dossat, A. M., Strong, C. E., Francis, T. C., et al. (2015). Methyl supplementation attenuates cocaine-seeking behaviors and cocaine-induced c-Fos activation in a DNA methylation-dependent manner. *Journal of Neuroscience, 35*(23), 8948–8958. http://dx.doi.org/10.1523/JNEUROSCI.5227-14.2015.

Wu, H., Wu, X., Shen, L., & Zhang, Y. (2014). Single-base resolution analysis of active DNA demethylation using methylase-assisted bisulfite sequencing. *Nature Biotechnology, 32*(12), 1231–1240. http://dx.doi.org/10.1038/nbt.3073.

Wu, H., & Zhang, Y. (2011). Mechanisms and functions of Tet protein-mediated 5-methylcytosine oxidation. *Genes and Development, 25*(23), 2436–2452. http://dx.doi.org/10.1101/gad.179184.111.

Yu, M., Hon, G. C., Szulwach, K. E., Song, C. X., Zhang, L., Kim, A., et al. (2012). Base-resolution analysis of 5-hydroxymethylcytosine in the mammalian genome. *Cell, 149*(6), 1368–1380. http://dx.doi.org/10.1016/j.cell.2012.04.027.

Zhang, B., & Horvath, S. (2005). A general framework for weighted gene co-expression network analysis. *Statistical Applications in Genetics and Molecular Biology, 4*. http://dx.doi.org/10.2202/1544-6115.1128 (Article 17).

Zhang, H., Herman, A. I., Kranzler, H. R., Anton, R. F., Simen, A. A., & Gelernter, J. (2012). Hypermethylation of OPRM1 promoter region in European Americans with alcohol dependence. *Journal of Human Genetics, 57*(10), 670–675. http://dx.doi.org/10.1038/jhg.2012.98.

Zhang, H., Herman, A. I., Kranzler, H. R., Anton, R. F., Zhao, H., Zheng, W., et al. (2013). Array-based profiling of DNA methylation changes associated with alcohol dependence. *Alcoholism, Clinical and Experimental Research, 37*(Suppl. 1), E108–E115. http://dx.doi.org/10.1111/j.1530-0277.2012.01928.x.

Zhang, H., Wang, F., Kranzler, H. R., Yang, C., Xu, H., Wang, Z., et al. (2014). Identification of methylation quantitative trait loci (mQTLs) influencing promoter DNA methylation of alcohol dependence risk genes. *Human Genetics, 133*(9), 1093–1104. http://dx.doi.org/10.1007/s00439-014-1452-2.

Zhang, R., Miao, Q., Wang, C., Zhao, R., Li, W., Haile, C. N., et al. (2013). Genome-wide DNA methylation analysis in alcohol dependence. *Addiction Biology, 18*(2), 392–403. http://dx.doi.org/10.1111/adb.12037.

Zhang, R. R., Cui, Q. Y., Murai, K., Lim, Y. C., Smith, Z. D., Jin, S., et al. (2013). Tet1 regulates adult hippocampal neurogenesis and cognition. *Cell Stem Cell, 13*(2), 237–245. http://dx.doi.org/10.1016/j.stem.2013.05.006, pii:S1934-5909(13)00199-9.

Zhang, X., Kusumo, H., Sakharkar, A. J., Pandey, S. C., & Guizzetti, M. (2014). Regulation of DNA methylation by ethanol induces tissue plasminogen activator expression in astrocytes. *Journal of Neurochemistry, 128*(3), 344–349. http://dx.doi.org/10.1111/jnc.12465.

Zhou, Z., Yuan, Q., Mash, D. C., & Goldman, D. (2011). Substance-specific and shared transcription and epigenetic changes in the human hippocampus chronically exposed to cocaine and alcohol. *Proceedings of the National Academy of Sciences of the United States of America, 108*(16), 6626–6631. http://dx.doi.org/10.1073/pnas.1018514108.

CHAPTER 9

What Does the Future Hold for the Study of Nucleic Acid Modifications in the Brain?

P.R. Marshall[1], T.W. Bredy[1,2]

[1]The University of California Irvine, Irvine, CA, United States; [2]The University of Queensland, Brisbane, QLD, Australia

Over the past 40 years, technological advances have made it possible to interrogate the entire genome; thus, the understanding of DNA modification has evolved significantly. Once considered to be a relatively static epigenetic mechanism, with its primary function restricted to the regulation of transcriptional programming during early cellular development, we now know that DNA methylation is a highly dynamic process in postmitotic neurons and plays a particularly important role in neuronal gene expression that directly impacts behavior. For example, in the adult brain, neuronal activity–induced changes in 5-methylcytosine (5mC) frequently occur outside gene promoters (Guo et al., 2011; Guo, Su, et al., 2011), and 5-hydroxymethylcytosine (5hmC) accounts for almost half of DNA methylation detected in the brain (Szulwach et al., 2011). Moreover, as is discussed herein, the base sequence can also dictate the relative probability that a region of the genome will be epigenetically modified. This relationship is important because it suggests that gene–epigenetic interactions should be considered in the context of a multilevel and bidirectional landscape, with other epigenetic regulators also acting to coordinate the function of the genome in a cell- and context-specific manner.

This correlation is particularly apparent in the discussions in Chapter 5 on non–CpG methylation and the functionally distinct role of 5hmC in the brain. However, despite this evidence much of the field still assumes that there is an inverse correlation between 5mC in gene promoters and gene expression. This leads us to ask what direction the field now needs to take. It is proposed that we will soon come to realize that it does not take millennia for DNA to change its function. Rather, mounting evidence suggests that DNA is constantly evolving and adapting in real time to its environment and that epigenetic modifications and related downstream effects on gene regulation rapidly contribute to phenotypic diversity from the level of cell function up to behavior. In the following sections, we present some of the most interesting molecular mechanisms related to genome regulation and how they are subject to DNA modification. It is evident that we have scratched only the tip of the iceberg with respect to this important mechanism of gene–environment interaction.

DNA Modifications in the Brain
ISBN 978-0-12-801596-4
http://dx.doi.org/10.1016/B978-0-12-801596-4.00009-5

EXPANDING THE REALM OF POSSIBILITY: DNA MODIFICATIONS ON ALL FOUR BASES

To date, all published research aimed at elucidating the role of DNA modifications in the mammalian brain has focused on either 5mC or the rediscovered 5hmC, which is a functionally distinct oxidative derivative of 5mC (Baker-Andresen, Ratnu, & Bredy, 2013; Lister et al., 2013). 5mC and 5hmC are highly prevalent in neurons relative to other cell types, and both modifications are regulated in response to experience (Li, Baker-Andresen, Zhao, Marshall, Bredy, 2014; Li et al., 2014; Miller & Sweatt, 2007). Although more than 20 DNA modifications have been identified, the majority that have been found in eukaryotes have only been studied in the context of DNA lesion and repair (Korlach & Turner, 2012), with relatively nothing being known about their role in regulating brain plasticity. This limited appreciation of what may be a highly complex and fundamental process has been the result of the relatively low frequency at which these marks have been detected using standard biochemical assays. It is plausible that, due to technical limitations, the important contribution of this vast and complex epigenetic regulatory mechanism toward experience-dependent gene expression has been largely overlooked.

With the advent of new sequencing and chemical-based nucleotide tagging approaches (as described in Chapter 8 and in the following), signal detection issues are being resolved and accurate genome-wide characterization of novel base modifications will soon be achieved. This is where chemistry and neurobiology have merged to achieve remarkably rapid advances in the past 5 years. For example, novel chemical tagging has led to the discovery of N^6-methyldeoxyadenosine (m6dA) in the eukaryotic genome. This epigenetic mark had all but been written off as a rare base lesion in the 1990s; however, recent genome-wide profiling studies have led to the surprising discovery that in *Chlamydomonas reindardtii*, m6dA accumulates at transcription start sites (Fu et al., 2015) and its level increases across development and is enriched within transposable elements in *Drosophila* (Zhang et al., 2015). Furthermore, m6dA seems to be involved in reproductive viability in *Caenorhabditis elegans* (Greer et al., 2015). Koziol et al. (2016) have found pervasive accumulation of m6dA throughout the vertebrate genome.

Similarly, in an unpublished series of experiments, we have discovered that m6dA is also a prevalent mark in neurons of the mammalian brain and that m6dA deposition is localized to the promoter of an activity-inducible gene known to be critical for memory formation. Furthermore, we have identified a putative N^6-adenine–specific methyltransferase, N6amt1, as a key enzyme responsible for the inducible deposition of N^6-methyladenine (m6A) in postmitotic neurons of the adult brain. This base modification has also been shown to engage mechanisms of base excision repair (Franchini et al., 2014), which is very intriguing in light of evidence indicating that DNA double-strand breaks are activity dependent and that complementary DNA repair pathways are induced within the adult cortex in response to learning (Madabhushi et al., 2015). Thus, it is evident that the information processing capacity of DNA is far more complex than currently appreciated, and the finding that m6dA is a potent regulator of activity-induced gene expression is a major advance in the field of neuroepigenetics.

Regardless of these interesting new threads, the central question of why so many DNA base modifications would exist if they are invoked primarily as a function of oxidative stress or in response to ionizing radiation or other mutagenic stimuli in our environment, remains to be addressed. Furthermore, the field has also remained stagnant regarding the role of DNA since Watson and Crick first proposed the central dogma in the 1950s, which has led to an oversimplified view of DNA as a *static* carrier of information. DNA is more than just a heritable series of four letter variations. It is a dynamic, living chemistry that is the source and endpoint of adaptation and the selective pressures of evolution. For example, although DNA can control its own packaging in chromatin and regulate transcription via sequence-related properties (Ernst et al., 2011; Ernst Komor, Barton, 2011), its function is also influenced through dynamic changes to its *structure and physical chemistry* (Parker, Hansen, Abaan, Tullius, & Margulies, 2009; Rohs et al., 2009; Rohs, West, Liu, Honig, 2009). Moreover, DNA flexibility and RNA stability are highly correlated with dinucleotide sequence composition (Heddi, Abi-Ghanem, Lavigne, & Hartmann, 2010) and, as discussed in the next section, each of these factors can be influenced by DNA modification.

DNA MODIFICATIONS DRIVE DNA STRUCTURE AND FUNCTION

Since Watson and Crick proposed the model of the right-handed double helix, which was later called B-DNA, the field has come to appreciate that this particular conformational state is but one possible shape. In fact, when the crystal structure of DNA was revealed, the field was amazed to discover that DNA could also assume a zig-zag confirmation with a left-handed double helix, which was completely opposite to that proposed 25 years earlier (Wang et al., 1979). It is evident that this interesting conformational state of DNA, or Z-DNA as it became known, together with at least 20 other possible DNA states such as the G-quadruplex, which interacts with DNA helicases and other epigenetic regulators, have very important biological functions in the cell (Murat & Balasubramanian, 2014; Rich & Zhang, 2003). It is now known that the structural properties of DNA can influence its ability to recognize and interact with transcription factors and other protein partners that, in turn, drive gene expression as well as coordinating the organization and integrity of the genome (Parker et al., 2009; Rohs et al., 2009; Rohs, West, et al., 2009; H. Zhou et al., 2015; K.I. Zhou et al., 2015; and reviewed in Harteis & Schneider, 2014).

Perhaps most important for this discussion, the structural reactivity and conformational state of DNA is profoundly influenced by base modifications. DNA methylation decreases the flexibility of DNA to interfere with the exaggerated bending of DNA required to form nucleosomes (Nathan & Crothers, 2002), resulting in a further shortening of the regions of linker DNA (Choy et al., 2010) that can alter the conformational space of a gene (Bettecken, Frenkel, & Trifonov, 2011). Specifically, the accumulation of the cytosine modification 5-formylcytosine (5fC) results in the formation of a DNA structure called F-DNA, which is characterized by helical underwinding (Raiber et al., 2015). This modification is thought to impact DNA supercoiling as well as the packaging of DNA into chromatin, both of which are related to transcription. Moreover, 5hmC inhibits the transition from B- to Z-DNA, whereas the oxidized derivatives 5fC and 5-carboxylcytosine seem to promote this process (Nickol, Behe, & Felsenfeld, 1982; Wang et al., 2014). Interestingly, Z-DNA occurs predominantly in a CpG dinucleotide context within repetitive elements across the genome (ie, CpG islands). Z-DNA is also found near promoter regions and, like F-DNA, stimulates gene transcription (Oh, Kim, Rich, 2002; Rich & Zhang, 2003). A hallmark feature of G-quadruplex DNA is that it tends to occur in hypomethylated regions, and it is associated with genome instability and DNA damage (De & Michor, 2011).

DNA MODIFICATIONS INFLUENCE DNA EDITING

Beyond the well-characterized process V(D)J recombination and somatic hypermutation associated with diversification of the immune system, there are emerging reports of other forms of functionally relevant DNA editing, including retrotransposon insertion and dynamic single-nucleotide variants (SNVs), which can occur anywhere in the neuronal genome. Although it was previously thought that retrotransposition occurred primarily during early embryogenesis (Kano et al., 2009), it has now been demonstrated that the expression of the long interspersed nuclear element 1 (L1) continues during adulthood and is elevated in the brain (Muotri et al., 2010). L1 retrotransposition occurs in response to a range of environmental stimuli, including voluntary exercise and chronic cocaine exposure (Muotri, Zhao, Marchetto, & Gage, 2009; Maze et al., 2011), and it has recently been proposed to generate the unique experience-dependent transcriptome profile of individual neurons. Indeed, single-cell retrotransposon sequencing analysis has revealed pervasive L1 mobilization in human hippocampal neurons (Upton et al., 2015). L1 transcription is repressed by region-specific DNA methylation in the 5′ untranslated region (Hata & Sakaki, 1997) and, accordingly, in neural precursors and differentiated neurons, by the expression of methyl-CpG-binding protein 2 (Yu et al., 2001). Furthermore, L1 elements undergo an age-related depletion of 5hmC in the hippocampus (Szulwach et al., 2011), which may reflect reduced plasticity. The insertion or deletion of transposable elements, which is heavily determined by DNA modifications, influences gene expression by introducing novel alternative promoter regions, enhancer elements, and transcription factor binding sites or by promoting the formation of heterochromatin (Feschotte, 2008).

With respect to SNVs, it is becoming increasingly evident that each postmitotic neuron in the human cortex can have a distinct genome, with conservative estimates of approximately 1500 somatic SNVs per neuron (Lodato et al., 2015), whereas others suggest up to 10,000 SNVs may accumulate in healthy differentiated neurons across the life span (Hazen et al., 2016). What is particularly intriguing about these observations is that the majority of edits occur in noncoding regions proximal to transcriptionally active protein-coding genes that are critical for neuronal function, and seem to be the products of deamination (ie, 5mC-T conversion). Indeed, DNA deamination serves as a major mechanism of DNA editing (Nabel, Manning, Kohli, 2012; Nabel et al., 2012), itself a function of modified DNA, the relevance of which has yet to be appreciated. Sandra Pena de Ortiz and colleagues have, for many years, reported on the profound endonuclease activity that occurs in the brain in response to experience (Castro-Pérez et al., 2016; Colón-Cesario et al., 2006; Saavedra-Rodríguez et al., 2009). Together, the evidence suggests that DNA editing events driven by covalent modifications may serve as a critically important source of functional diversification in postmitotic neurons, enabling them to optimize their transcriptional responses to rapidly changing environmental signals. This concept is a significant departure from DNA as a *static* carrier of heritable information.

TAKING NEUROEPIGENETICS TO THE NEXT LEVEL: EPITRANSCRIPTOMICS COMES OF AGE

It has been known for at least half a century that, like DNA, RNA is also subject to chemical modification, with more than 140 marks identified to date (Machnicka et al., 2013). These posttranscriptional "epitranscriptomic" modifications, which direct the functional readout of nascent RNAs in a highly structured and coordinated manner, have now been found to occur on all classes of RNA, including mRNA, as well as small and long noncoding RNAs (Saletore et al., 2012). For example, depending on the locus, chemical modifications to RNA dictate patterns of alternative splicing and degradation (Liu et al., 2016), influence secondary structure (Spitale et al., 2015), and impact the rate of translation (H. Zhou et al., 2015; K.I. Zhou et al., 2015). Thus, RNA modification may serve as yet another epigenetic "code" for the regulation of activity-dependent changes in RNA, imparting diversity to its function without the need for further increased levels of transcription.

One RNA modification, m6A, of which there are several readers and writers (Roundtree & He, 2016), is highly abundant throughout the mammalian transcriptome and seems to be involved in a variety of biological processes (Dominissini et al., 2012;

Fu, Dominissini, Rechavi, & He, 2014; Meyer & Jaffrey, 2014; Meyer et al., 2012). In the mouse brain, RNA m6A is developmentally regulated and increases in adulthood (Meyer et al., 2012), which suggests a potential role in posttranscriptional regulation of RNA associated with neural plasticity and behavioral adaptation. Zhou et al. (2015) reported the activity-dependent nature of mammalian m6A was demonstrated in response to heat shock stress. In unpublished work, we have discovered that RNA m6A is also dynamic in the mouse brain in response to behavioral training, which is reflected by widespread learning-induced, locus-specific accumulation of this epitranscriptomic mark. Importantly, when the RNA methylation is amplified after knockdown of the RNA demethylase FTO, memory is enhanced.

In addition, RNA methylation can target mRNAs for degradation and, depending on the context, promote their translation (Dominissini et al., 2016; Wang et al., 2015; H. Zhou et al., 2015; K.I. Zhou et al., 2015). This process occurs in the mammalian brain in a dynamic and experience-dependent manner and may therefore provide a putative explanation for the paradoxical relationship between experience-induced mRNA and protein levels in the brain that is so often observed. We predict that, like DNA modification, RNA modification will come to be appreciated as a critically important epigenetic modification associated with behavioral adaptation. The future is very bright for epitranscriptomics.

OUTSTANDING QUESTIONS AND EMERGING TECHNOLOGIES

DNA modifications are involved in synaptic plasticity and adaptation in the adult brain (Guo et al., 2011; Guo, Su, Zhong, Ming, Song, 2011; Kaas et al., 2013; Li, Baker-Andresen, et al., 2014; Li et al., 2014; Ma et al., 2009; Matrisciano, Dong, Gavin, Nicoletti, & Guidotti, 2011; Rudenko et al., 2013); however, it remains unclear how this epigenetic mechanism is regulated in a cell type- and region-specific manner. Do activated and quiescent neurons exhibit similar epigenomic profiles or can they be distinguished in response to experience? Active DNA demethylation involves oxidation, deamination, and DNA repair; are these enzyme-mediated modifications of DNA interconnected or do they act independently during the formation of memory? DNA methylation occurs at different genomic loci within different cells (Wu et al., 2010). Given that DNA modifications are abundant throughout the neuronal genome, does this epigenetic regulatory network serve different functions depending on where it accumulates in response to experience? Indeed, non–CpG DNA methylation has been demonstrated to be involved in regulating gene expression within the mammalian brain (Xie et al., 2012), and, with respect to 5hmC, Ten-eleven translocation (Tet) proteins are known to bind to non–CpG methylated loci (Xu et al., 2012). However, direct evidence for non–CpG hydroxymethylation or other base modifications within the brain, and whether they are sensitive to experience, remains elusive. Future studies will require advances in cell type- and genomic locus-specific resolution. To detect differences in active DNA demethylation in specific cell types, fluorescence-activated cell sorting (FACS) technology can be used. For example, by targeting a neuronal-specific marker NeuN, FACS can enrich for neurons from homogenized cell populations in the adult brain (Guez-Barber et al., 2012). Moreover, in combination with other markers, this approach can be used to determine specific patterns of active DNA demethylation within excitatory neurons, inhibitory neurons, astrocytes and glia, and how cell-specific demethylation pathways interconnect to regulate synaptic activity.

One of the most revolutionary technical advances in recent years, bioorthogonal click chemistry is poised to provide an unprecedented level of detail in our understanding of nucleotide modifications and their effect on gene expression in the brain. This approach uses an extremely efficient azide-alkyne cycloaddition reaction, which is catalyzed by copper (CuAAC), to attach biotin handles to and therefore enrich for specific DNA modifications. Coupled with high-throughput sequencing, click DNA labeling, as it is commonly known, enables genome-wide profiling of any DNA modification at single-base resolution (Li & Chen, 2016). Newer approaches using copper-free reactions have since enabled the profiling of modified bases in live cells, zebrafish, and mice (Sletten & Bertozzi, 2011). For example, Tet1-associated bisulfite sequencing permits the determination of 5hmC (mediated by Tet1) at single-nucleotide resolution. This method allows researchers to analyze DNA demethylation on any individual cytosine across the genome (Yu et al., 2012). Perhaps even more remarkable, this technology is now being integrated with optical probes and small molecule–based methods to enable noninvasive, photoinducible tagging of nucleotides. This approach will most likely play itself out in the realm of RNA epigenetics, where it is rapidly becoming clear that the many chemical modifications known to occur on RNA will impact its function in subtle, but powerful ways (ie, through RNA modification–induced changes in nascent RNA structure and compartmentalization in the cell, Kubota, Tran, & Spitale, 2015; Spitale et al., 2015).

With the exceptional level of detail in the measurement of underlying epigenetic changes on the horizon via datasets produced using chemispecific, activity-enriched, cell type–specific methods, this begs the question as to what is required to best use the information generated from these studies. Fortunately, an emerging suite of techniques is being developed to locus-specifically edit and direct putative DNA modifications in the brain, namely, the clustered regularly interspaced palindromic repeat-Cas9 and transcriptional activator-like effector technology. These approaches use the enzymatic machinery from bacteria, which normally function to target and destroy viruses and exchange the RNA targeting sequence for one of an experimenter's choice to selectively target regions of DNA for modification (Doudna & Charpentier, 2014; Hsu, Lander, & Zhang, 2014). Furthermore, and more excitingly for neuroepigenetics, this technique can also be modified to use a deactivated version of the cutting enzyme (dubbed dCas9), to site selectively direct proteins, or epigenetic modifiers to intact loci in awake behaving animals.

This provides the ability to elegantly probe learning and memory processes and to draw more precise mechanistic conclusions. For example, Heller et al. (2014) have used this approach to show the sufficiency of modifications around one locus to modulate a behavior related to reward seeking.

SUMMARY AND CONCLUSIONS

As we expand the notion of base modifications and delve deeper into the issue of their functional relevance in the brain, a reinterpretation of the concept of the gene may be required. Based on what we are learning about neuroepigenetic mechanisms, this reinterpretation will likely involve a more holistic definition of what constitutes mutation and epimutation (Babbitt, Coppola, Alawad, Hudson, 2016), and an appreciation that what rules hold for gene programming in early development do not necessarily apply in postmitotic neurons of the adult brain. A detailed characterization of the impact of all DNA modifications on the molecular mechanisms controlling experience-dependent gene expression will need to be performed across development and in a brain region– and cell type–specific manner. With new NIH-supported initiatives coming online, including the 3D Nucleosome and the 4D Transcriptome projects, efforts are underway to move in this direction. Thus, a deeper understanding of the fundamental nature of nucleic acids and their chemically modified derivatives in the brain, as well as insight regarding the information that they are capable of transmitting in response to environmental cures, is imminent. In particular, we predict that a large number of functional modifications on *all four* canonical nucleobases remain to be discovered, and it will be within the realm of memory and behavioral adaptation where the impact of these novel epigenetic purveyors of genomic and behavioral diversity will be most significant.

REFERENCES

Babbitt, G. A., Coppola, E. E., Alawad, M. A., & Hudson, A. O. (2016). Can all heritable biology really be reduced to a single dimension? *Gene, 578*, 162–168. http://dx.doi.org/10.1016/j.gene.2015.12.043.

Baker-Andresen, D., Ratnu, V. S., & Bredy, T. W. (2013). Dynamic DNA methylation: a prime candidate for genomic metaplasticity and behavioral adaptation. *Trends in Neurosciences, 36*, 3–13. http://dx.doi.org/10.1016/j.tins.2012.09.003.

Bettecken, T., Frenkel, Z. M., & Trifonov, E. N. (2011). Human nucleosomes: special role of CG dinucleotides and Alu-nucleosomes. *BMC Genomics, 12*, 273. http://dx.doi.org/10.1186/1471-2164-12-273.

Castro-Perez, E., et al. (2016). Identification and characterization of the V(D)J recombination activating gene 1 in long-term memory of context fear conditioning. *Neural Plasticity, 2016*, 1752176. http://dx.doi.org/10.1155/2016/1752176.

Choy, J. S., et al. (2010). DNA methylation increases nucleosome compaction and rigidity. *Journal of the American Chemical Society, 132*, 1782–1783. http://dx.doi.org/10.1021/ja910264z.

Colon-Cesario, M., et al. (2006). An inhibitor of DNA recombination blocks memory consolidation, but not reconsolidation, in context fear conditioning. *The Journal of Neuroscience: the Official Journal of the Society for Neuroscience, 26*, 5524–5533. http://dx.doi.org/10.1523/JNEUROSCI.3050-05.2006.

De, S., & Michor, F. (2011). DNA secondary structures and epigenetic determinants of cancer genome evolution. *Nature Structural & Molecular Biology, 18*, 950–955. http://dx.doi.org/10.1038/nsmb.2089.

Dominissini, D., et al. (2012). Topology of the human and mouse m6A RNA methylomes revealed by m6A-seq. *Nature, 485,* 201–206. http://dx.doi.org/10.1038/nature11112.

Dominissini, D., et al. (2016). The dynamic N(1)-methyladenosine methylome in eukaryotic messenger RNA. *Nature, 530,* 441–446. http://dx.doi.org/10.1038/nature16998.

Doudna, J. A., & Charpentier, E. (2014). Genome editing. The new frontier of genome engineering with CRISPR-Cas9. *Science, 346,* 1258096. http://dx.doi.org/10.1126/science.1258096.

Ernst, J., et al. (2011). Mapping and analysis of chromatin state dynamics in nine human cell types. *Nature, 473,* 43–49. http://dx.doi.org/10.1038/nature09906.

Ernst, R. J., Komor, A. C., & Barton, J. K. (2011). Selective cytotoxicity of rhodium metalloinsertors in mismatch repair-deficient cells. *Biochemistry, 50,* 10919–10928. http://dx.doi.org/10.1021/bi2015822.

Feschotte, C. (2008). Transposable elements and the evolution of regulatory networks. *Nature Reviews. Genetics, 9,* 397–405. http://dx.doi.org/10.1038/nrg2337.

Franchini, D. M., & Petersen-Mahrt, S. K. (2014). AID and APOBEC deaminases: balancing DNA damage in epigenetics and immunity. *Epigenomics, 6,* 427–443. http://dx.doi.org/10.2217/epi.14.35.

Franchini, D. M., et al. (2014). Processive DNA demethylation via DNA deaminase-induced lesion resolution. *PLoS One, 9,* e97754. http://dx.doi.org/10.1371/journal.pone.0097754.

Fu, Y., Dominissini, D., Rechavi, G., & He, C. (2014). Gene expression regulation mediated through reversible m(6)A RNA methylation. *Nature Reviews. Genetics, 15,* 293–306. http://dx.doi.org/10.1038/nrg3724.

Fu, Y., Dominissini, D., Rechavi, G., & He, C., et al. (2015). N6-methyldeoxyadenosine marks active transcription start sites in chlamydomonas. *Cell, 161,* 879–892. http://dx.doi.org/10.1016/j.cell.2015.04.010.

Greer, E. L., et al. (2015). Dna methylation on N6-Adenine in *C. elegans. Cell, 161,* 868–878. http://dx.doi.org/10.1016/j.cell.2015.04.005.

Guez-Barber, D., et al. (2012). FACS purification of immunolabeled cell types from adult rat brainC. elegans. *Journal of Neuroscience Methods, 203,* 10–18. http://dx.doi.org/10.1016/j.jneumeth.2011.08.045.

Guo, J. U., et al. (2011). Neuronal activity modifies the DNA methylation landscape in the adult brain. *Nature Neuroscience, 14,* 1345–1351. http://dx.doi.org/10.1038/nn.2900.

Guo, J. U., Su, Y., Zhong, C., Ming, G. L., & Song, H. (2011). Hydroxylation of 5-methylcytosine by TET1 promotes active DNA demethylation in the adult brain. *Cell, 145,* 423–434. http://dx.doi.org/10.1016/j.cell.2011.03.022.

Harteis, S., Su, Y., Zhong, C., Ming, G. L., & Schneider, S. (2014). Making the bend: DNA tertiary structure and protein–DNA interactions. *International Journal of Molecular Sciences, 15,* 12335–12363. http://dx.doi.org/10.3390/ijms150712335.

Hata, K., & Sakaki, Y. (1997). Identification of critical CpG sites for repression of L1 transcription by DNA methylation. *Gene, 189,* 227–234. http://dx.doi.org/10.3390/ijms150712335.

Hazen, J. L., Faust, G. G., Rodriguez, A. R., Ferguson, W. C., Shumilina, S., Clark, R. A., et al. (March 16, 2016). The complete genome sequences, unique mutational spectra, and developmental potency of adult neurons revealed by cloning. *Neuron, 89*(6), 1223–1236. http://dx.doi.org/10.1016/j.neuron.2016.02.004.

Heddi, B., & Abi-Ghanem, J., Lavigne, M., & Hartmann, B. (2010). Sequence-dependent DNA flexibility mediates DNase I cleavage. *Journal of Molecular Biology, 395,* 123–133. http://dx.doi.org/10.1016/j.jmb.2009.10.023.

Heller, E. A., Abi-Ghanem, J., Lavigne, M., & Hartmann, B., et al. (2014). Locus-specific epigenetic remodeling controls addiction- and depression-related behaviors. *Nature Neuroscience, 17,* 1720–1727. http://dx.doi.org/10.1038/nn.3871.

Hsu, P. D., Lander, E. S., & Zhang, F. (2014). Development and applications of CRISPR-Cas9 for genome engineering. *Cell, 157,* 1262–1278. http://dx.doi.org/10.1016/j.cell.2014.05.010.

Kaas, G. A., Lander, E. S., & Zhang, F., et al. (2013). TET1 controls CNS 5-methylcytosine hydroxylation, active DNA demethylation, gene transcription, and memory formation. *Neuron, 79,* 1086–1093. http://dx.doi.org/10.1016/j.neuron.2013.08.032.

Kano, H., et al. (2009). L1 retrotransposition occurs mainly in embryogenesis and creates somatic mosaicism. *Genes & Development, 23,* 1303–1312. http://dx.doi.org/10.1101/gad.1803909.

Korlach, J., & Turner, S. W. (2012). Going beyond five bases in DNA sequencing. *Current Opinion in Structural Biology, 22,* 251–261. http://dx.doi.org/10.1016/j.sbi.2012.04.002.

Koziol, M. J., & Turner, S. W., et al. (2016). Identification of methylated deoxyadenosines in vertebrates reveals diversity in DNA modifications. *Nature Structural & Molecular Biology, 23,* 24–30. http://dx.doi.org/10.1038/nsmb.3145.

Kubota, M. , Tran, C., & Spitale, R. C. (2015). Progress and challenges for chemical probing of RNA structure inside living cells. *Nature Chemical Biology, 11*, 933–941. http://dx.doi.org/10.1038/nchembio.1958.

Li, J., Tran, C., & Chen, P. R. (2016). Development and application of bond cleavage reactions in bioorthogonal chemistry. *Nature Chemical Biology, 12*, 129–137. http://dx.doi.org/10.1038/nchembio.2024.

Li, X., & Chen, P. R., et al. (2014). Neocortical Tet3-mediated accumulation of 5-hydroxymethylcytosine promotes rapid behavioral adaptation. *Proceedings of the National Academy of Sciences USA, 111*, 7120–7125. http://dx.doi.org/10.1073/pnas.1318906111.

Li, X., Baker-Andresen, D., Zhao, Q., Marshall, V., & Bredy, T. W. (2014). Methyl CpG binding domain ultra-sequencing: a novel method for identifying inter-individual and cell-type-specific variation in DNA methylation. *Genes, Brain, and Behavior, 13*, 721–731. http://dx.doi.org/10.1111/gbb.12150.

Lister, R., Baker-Andresen, D., Zhao, Q., Marshall, V., & Bredy, T. W., et al. (2013). Global epigenomic reconfiguration during mammalian brain development. *Science, 341*, 1237905. http://dx.doi.org/10.1126/science.1237905.

Liu, K., et al., (June 2016). Structural and functional characterization of the proteins responsible for N6-Methyladenosine modification and recognition, *Current Protein & Peptide Science, 17*(4), 306–318.

Lodato, M. A., et al. (2015). Somatic mutation in single human neurons tracks developmental and transcriptional history. *Science, 350*, 94–98. http://dx.doi.org/10.1126/science.aab1785.

Ma, D. K., et al. (2009). Neuronal activity-induced Gadd45b promotes epigenetic DNA demethylation and adult neurogenesis. *Science, 323*, 1074–1077. http://dx.doi.org/10.1126/science.1166859.

Machnicka, M. A., et al. (2013). MODOMICS: a database of RNA modification pathways–2013 update. *Nucleic Acids Research, 41*, D262–D267. http://dx.doi.org/10.1093/nar/gks1007.

Madabhushi, R., et al. (2015). Activity-induced DNA breaks govern the expression of neuronal early-response genes. *Cell, 161*, 1592–1605. http://dx.doi.org/10.1016/j.cell.2015.05.032.

Matrisciano, F., Dong, E., Gavin, D. P., Nicoletti, F., & Guidotti, A. (2011). Activation of group II metabotropic glutamate receptors promotes DNA demethylation in the mouse brain. *Molecular Pharmacology, 80*, 174–182. http://dx.doi.org/10.1124/mol.110.070896.

Maze, I., Dong, E., Gavin, D. P., Nicoletti, F., & Guidotti, A., et al. (2011). Cocaine dynamically regulates heterochromatin and repetitive element unsilencing in nucleus accumbens. *Proceedings of the National Academy of Sciences USA, 108*, 3035–3040. http://dx.doi.org/10.1073/pnas.1015483108.

Meyer, K. D., & Jaffrey, S. R. (2014). The dynamic epitranscriptome: N6-methyladenosine and gene expression control. *Nature Reviews. Molecular Cell Biology, 15*, 313–326. http://dx.doi.org/10.1038/nrm3785.

Meyer, K. D., & Jaffrey, S. R., et al. (2012). Comprehensive analysis of mRNA methylation reveals enrichment in 3'UTRs and near stop codons. *Cell, 149*, 1635–1646. http://dx.doi.org/10.1016/j.cell.2012.05.003.

Miller, C. A., & Sweatt, J. D. (2007). Covalent modification of DNA regulates memory formation. *Neuron, 53*, 857–869. http://dx.doi.org/10.1016/j.neuron.2007.02.022.

Muotri, A. R., & Zhao, C. , Marchetto, M. C., & Gage, F. H. (2009). Environmental influence on L1 retrotransposons in the adult hippocampus. *Hippocampus, 19*, 1002–1007. http://dx.doi.org/10.1002/hipo.20564.

Muotri, A. R., Zhao, C., Marchetto, M. C., & Gage, F. H., et al. (2010). L1 retrotransposition in neurons is modulated by MeCP2. *Nature, 468*, 443–446. http://dx.doi.org/10.1038/nature09544.

Murat, P. , & Balasubramanian, S. (2014). Existence and consequences of G-quadruplex structures in DNA. *Current Opinion in Genetics & Development, 25*, 22–29. http://dx.doi.org/10.1016/j.gde.2013.10.012.

Nabel, C. S., & Balasubramanian, S., et al. (2012). AID/APOBEC deaminases disfavor modified cytosines implicated in DNA demethylation. *Nature Chemical Biology, 8*, 751–758. http://dx.doi.org/10.1038/nchembio.1042.

Nabel, C. S., Manning, S. A., & Kohli, R. M. (2012). The curious chemical biology of cytosine: deamination, methylation, and oxidation as modulators of genomic potential. *ACS Chemical Biology, 7*, 20–30. http://dx.doi.org/10.1021/cb2002895.

Nathan, D. , Manning, S. A., & Crothers, D. M. (2002). Bending and flexibility of methylated and unmethylated EcoRI DNA. *Journal of Molecular Biology, 316*, 7–17. http://dx.doi.org/10.1006/jmbi.2001.5247.

Nickol, J., & Behe, M. , & Felsenfeld, G. (1982). Effect of the B–Z transition in poly(dG-m5dC). poly(dG-m5dC) on nucleosome formation. *Proceedings of the National Academy of Sciences USA, 79*, 1771–1775. http://dx.doi.org/10.1006/jmbi.2001.5247.

Oh, D. B., Kim, Y. G., & Rich, A. (2002). Z-DNA-binding proteins can act as potent effectors of gene expression in vivo. *Proceedings of the National Academy of Sciences USA, 99*, 16666–16671. http://dx.doi.org/10.1073/pnas.262672699.

Parker, S. C., Hansen, L. , & Abaan, H. O., Tullius, T. D., & Margulies, E. H. (2009). Local DNA topography correlates with functional noncoding regions of the human genome. *Science*, *324*, 389–392. http://dx.doi.org/10.1126/science.1169050.

Raiber, E. A., Hansen, L., Abaan, H. O., Tullius, T. D., & Margulies, E. H., et al. (2015). 5-Formylcytosine alters the structure of the DNA double helix. *Nature Structural & Molecular Biology*, *22*, 44–49. http://dx.doi.org/10.1038/nsmb.2936.

Rich, A. , & Zhang, S. (2003). Timeline: Z-DNA: the long road to biological function. *Nature Reviews. Genetics*, *4*, 566–572. http://dx.doi.org/10.1038/nrg1115.

Rohs, R., & Zhang, S., et al. (2009). The role of DNA shape in protein-DNA recognition. *Nature*, *461*, 1248–1253. http://dx.doi.org/10.1038/nature08473.

Rohs, R., West, S. M., Liu, P., & Honig, B. (2009). Nuance in the double-helix and its role in protein-DNA recognition. *Current Opinion in Structural Biology*, *19*, 171–177. http://dx.doi.org/10.1016/j.sbi.2009.03.002.

Roundtree, I. A., West, S. M., Liu, P., & He, C. (2016). RNA epigenetics-chemical messages for post-transcriptional gene regulation. *Current Opinion in Chemical Biology*, *30*, 46–51. http://dx.doi.org/10.1016/j.cbpa.2015.10.024.

Rudenko, A. , & He, C., et al. (2013). Tet1 is critical for neuronal activity-regulated gene expression and memory extinction. *Neuron*, *79*, 1109–1122. http://dx.doi.org/10.1016/j.neuron.2013.08.003.

Saavedra-Rodriguez, L., et al. (2009). Identification of flap structure-specific endonuclease 1 as a factor involved in long-term memory formation of aversive learning. *The Journal of Neuroscience: The Official Journal of the Society for Neuroscience*, *29*, 5726–5737. http://dx.doi.org/10.1523/JNEUROSCI.4033-08.2009.

Saletore, Y., et al. (2012). The birth of the Epitranscriptome: deciphering the function of RNA modifications. *Genome Biology*, *13*, 175–5737. http://dx.doi.org/10.1186/gb-2012-13-10-175.

Sletten, E. M., & Bertozzi, C. R. (2011). From mechanism to mouse: a tale of two bioorthogonal reactions. *Accounts of Chemical Research*, *44*, 666–676. http://dx.doi.org/10.1021/ar200148z.

Spitale, R. C., & Bertozzi, C. R., et al. (2015). Structural imprints in vivo decode RNA regulatory mechanisms. *Nature*, *519*, 486–490. http://dx.doi.org/10.1038/nature14263.

Szulwach, K. E., et al. (2011). 5-hmC-mediated epigenetic dynamics during postnatal neurodevelopment and aging. *Nature Neuroscience*, *14*, 1607–1616. http://dx.doi.org/10.1038/nn.2959.

Upton, K. R., et al. (2015). Ubiquitous L1 mosaicism in hippocampal neurons. *Cell*, *161*, 228–239. http://dx.doi.org/10.1016/j.cell.2015.03.026.

Wang, A. H., et al. (1979). Molecular structure of a left-handed double helical DNA fragment at atomic resolution. *Nature*, *282*, 680–686. http://dx.doi.org/10.1016/j.cell.2015.03.026.

Wang, S. , et al. (2014). Systematic investigations of different cytosine modifications on CpG dinucleotide sequences: the effects on the B-Z transition. *Journal of the American Chemical Society*, *136*, 56–59. http://dx.doi.org/10.1021/ja4107012.

Wang, X., et al. (2015). N(6)-methyladenosine modulates messenger RNA translation efficiency. *Cell*, *161*, 1388–1399. http://dx.doi.org/10.1016/j.cell.2015.05.014.

Wu, H., et al. (2010). Dnmt3a-dependent nonpromoter DNA methylation facilitates transcription of neurogenic genes. *Science*, *329*, 444–448. http://dx.doi.org/10.1126/science.1190485.

Xie, W., et al. (2012). Base-resolution analyses of sequence and parent-of-origin dependent DNA methylation in the mouse genome. *Cell*, *148*, 816–831. http://dx.doi.org/10.1016/j.cell.2011.12.035.

Xu, Y., et al. (2012). Tet3 CXXC domain and dioxygenase activity cooperatively regulate key genes for Xenopus eye and neural development. *Cell*, *151*, 1200–1213. http://dx.doi.org/10.1016/j.cell.2012.11.014.

Yu, M., et al. (2012). Tet-assisted bisulfite sequencing of 5-hydroxymethylcytosine. *Nature Protocols*, *7*, 2159–2170. http://dx.doi.org/10.1038/nprot.2012.137.

Yu, F., Zingler, N., Schumann, G., & Strätling, W. H. (November 1, 2001). Methyl-CpG-binding protein 2 represses LINE-1 expression and retrotransposition but not Alu transcription. *Nucleic Acids Research*, *29*(21), 4493–4501.

Zhang, G., et al. (2015). N6-methyladenine DNA modification in *Drosophila*. *Cell*, *161*, 893–906. http://dx.doi.org/10.1016/j.cell.2015.04.018.

Zhou, H., et al. (2015). New insights into Hoogsteen base pairs in DNA duplexes from a structure-based survey Drosophila. *Nucleic Acids Research*, *43*, 3420–3433. http://dx.doi.org/10.1093/nar/gkv241.

Zhou, K. I., et al. (2015). N-Methyladenosine modification in a long noncoding RNA Hairpin Predisposes its conformation to protein binding. *Journal of Molecular Biology*, *43*. , 3420–3433. http://dx.doi.org/10.1016/j.jmb.2015.08.021.

INDEX

'Note: Page numbers followed by "f" indicate figures, "t" indicate tables.'